A HUNDRED AND SEVENTY CHINESE POEMS

A HUNDRED AND SEVENTY CHINESE POEMS

TRANSLATED BY

ARTHUR WALEY

ISBN: 978-93-5128-834-3(PB)

First Published in 1918
Indian Reprint in 2017

Published by

Kalpaz Publications
C-30, Satyawati Nagar,
Delhi – 110052
E-mail: kalpaz@hotmail.com
Website: kalpazpublications.com
Ph.: +91-9212142040

Cataloging in Publication Data—DK
Courtesy: D.K. Agencies (P) Ltd. <docinfo@dkagencies.com>

A hundred and seventy Chinese poems / translated by Arthur Waley.
pages cm
"First published in 1918"—Title page verso.
ISBN 9789351288343

1. Chinese poetry—Translations into English. I. Waley, Arthur, translator.

PL2658.E3H86 2017 DDC 895.11008 23

ERRATA

P. 21, heading, for BIOGRAPHICAL read BIBLIOGRAPHICAL.
P. 27, l. 7 „ *single of* „ *single one of.*
P. 29, l. 9 „ *eat* „ *ate.*
P. 32, l. 23 „ *houses* „ *house.*
P. 65, l. 3 „ *standing* „ *stand.*
P. 88, l. 15 „ *pillar* „ *pillow.*
P. 109, l. 22 „ *Memories* „ *Mémoires.*
P. 116, last line „ *Turn* „ *Turns.*
P. 134, l. 10 „ *and of Wu* „ *and Wu.*
P. 165, l. 13 „ *the things* „ *these things.*

PRELIMINARY NOTE

IN making this book I have tried to avoid poems which have been translated before. A hundred and forty of those I have chosen have not been translated by any one else. The remaining thirty odd I have included in many cases because the previous versions were full of mistakes; in others, because the works in which they appeared are no longer procurable. Moreover, they are mostly in German, a language with which my readers may not all be acquainted.

With some hesitation I have included literal versions of six poems (three of the "Seventeen Old Poems," "Autumn Wind," "Li Fu jēn," and "On the Death of his Father") already skilfully rhymed by Professor Giles in "Chinese Poetry in English Verse." They were too typical to omit; and a comparison of the two renderings may be of interest. Some of these translations have appeared in the "Bulletin of the School of Oriental Studies," in the "New Statesman," in the "Little Review" Chicago), and in "Poetry" (Chicago).

CONTENTS

PART I

PAGE

Introduction 3
The Method of Translation 19
Biographical Notes 21
Chapter I:
Battle 23
The Man-Wind and the Woman-Wind . . . 24
Master Tēng-t'u 26
The Orphan 27
The Sick Wife 29
Cock-Crow Song 30
The Golden Palace 31
"Old Poem". 32
Meeting in the Road 32
Fighting South of the Castle 33
The Eastern Gate 34
Old and New 35
South of the Great Sea 35
The Other Side of the Valley 36
Oaths of Friendship 37
Burial Songs 38
Seventeen Old Poems 39-48
The Autumn Wind 48
Li Fu-jēn 49
Song of Snow-white Heads 50

PAGE

CHAPTER I—*continued.*

To his Wife 51
Li Ling 52
Lament of Hsi-chün 53
Ch'in Chia 53
Ch'in Chia's Wife's Reply 54
Song 55

CHAPTER II:

Satire on Paying Calls in August 57
On the Death of his Father 58
The Campaign against Wu 59
The Ruins of Lo-yang 60
The Cock-fight 61
A Vision 62
The Curtain of the Wedding Bed 63
Regret 63
Taoist Song 64
A Gentle Wind 64
Woman 65
Day Dreams 66
The Scholar in the Narrow Street 66
The Desecration of the Han Tombs 67
Bearer's Song 68
The Valley Wind 69

CHAPTER III:

Poems by T'ao Ch'ien 71-79

CHAPTER IV:

Inviting Guests 81
Climbing a Mountain 81
Sailing Homeward 82
Five "Tzŭ-yeh" Songs 83

PAGE

CHAPTER IV—*continued.*

The Little Lady of Ch'ing-hsi 84
Plucking the Rushes 84
Ballad of the Western Island in the North Country . 84
Song 86
Song of the Men of Chin-ling 86
The Scholar Recruit 87
The Red Hills 87
Dreaming of a Dead Lady 88
The Liberator 89
Lo-yang 89
Winter Night 90
The Rejected Wife 90
People hide their Love 91
The Ferry 91
The Waters of Lung-t'ou 92
Flowers and Moonlight on the Spring River . . 92
Tchirek Song 93

CHAPTER V:

Business Men 95
Tell me now 95
On Going to a Tavern . . 96
Stone Fish Lake 96
Civilization 97
A Protest in the Sixth Year of Ch'ien Fu . 97
On the Birth of his Son 98
The Pedlar of Spells 98
Boating in Autumn 99
The Herd-boy 99
How I sailed on the Lake till I came to the Eastern Stream 100
A Seventeenth-century Chinese Poem 100

PART II

PAGE
INTRODUCTION 105
BY PO CHÜ-I:
An Early Levée 115
Being on Duty all night in the Palace and dreaming of the Hsien-yu Temple 116
Passing T'ien-mēn Street in Ch'ang-an and seeing a distant View of Chung-nan Mountain . . . 116
The Letter 117
Rejoicing at the Arrival of Ch'ēn Hsiung . . . 118
Golden Bells 119
Remembering Golden Bells 120
Illness 120
The Dragon of the Black Pool 121
The Grain-tribute 123
The People of Tao-chou . . . 123
The Old Harp 125
The Harper of Chao 125
The Flower Market . . 126
The Prisoner 127
The Chancellor's Gravel-drive . . 131
The Man who Dreamed of Fairies 132
Magic 134
The Two Red Towers 135
The Charcoal-seller . . . 137
The Politician 138
The Old Man with the Broken Arm 139
Kept waiting in the Boat at Chiu-k'ou Ten Days by an adverse Wind 142
On Board Ship: Reading Yüan Chēn's Poems . . 142
Arriving at Hsün-yang 143
Madly Singing in the Mountains 144

PAGE

By Po Chü-i—*continued.*

Releasing a migrant "Yen" (wild Goose) 145
To a Portrait Painter who desired him to sit . . . 146
Separation 147
Having climbed to the topmost Peak of the Incense-burner Mountain 148
Eating Bamboo-shoots 149
The Red Cockatoo 149
After Lunch 150
Alarm at first entering the Yang-tze Gorges . . . 150
On being removed from Hsün-yang and sent to Chung-chou 151
Planting Flowers on the Eastern Embankment . . 152
Children 153
Pruning Trees 154
Being visited by a Friend during Illness 155
On the way to Hangchow: Anchored on the River at Night 155
Stopping the Night at Jung-yang. 156
The Silver Spoon 156
The Hat given to the Poet by Li Chien 157
The Big Rug. 157
After getting Drunk, becoming Sober in the Night . 158
Realizing the Futility of Life 158
Rising Late and Playing with A-ts'ui, aged Two . 159
On a Box containing his own Works 160
On being Sixty 161
Climbing the Terrace of Kuan-yin and looking at the City 162
Climbing the Ling Ying Terrace and looking North . 162
Going to the Mountains with a little Dancing Girl, aged Fifteen 163

PAGE

BY PO CHÜ-I—*continued.*

Dreaming of Yüan Chēn 163
A Dream of Mountaineering 164
Ease 165
On hearing someone sing a Poem by Yüan Chēn . 165
The Philosophers 166
Taoism and Buddhism 167
Last Poem 168

PART I

INTRODUCTION

PRINCIPAL CHINESE DYNASTIES

Han, 206 B.C.—A.D. 220.
Wei, 220-264.
Chin, 265-419.
(Northern Wei, ruled over the North of China, 386-532.)
Liang, 502-556.
Sui, 589-618.
T'ang, 618-905.
Sung, 960-1278.
Yüan (Mongols), 1260-1341.
Ming, 1368-1640.
Ch'ing (Manchus), 1644-1912.

THE LIMITATIONS OF CHINESE LITERATURE

THOSE who wish to assure themselves that they will lose nothing by ignoring Chinese literature, often ask the question: "Have the Chinese a Homer, an Aeschylus, a Shakespeare or Tolstoy?" The answer must be that China has no epic and no dramatic literature of importance. The novel exists and has merits, but never became the instrument of great writers.

Her philosophic literature knows no mean between the traditionalism of Confucius and the nihilism of Chuang-tzŭ. In mind, as in body, the Chinese were for the most part torpid mainlanders. Their thoughts set out on no strange quests and adventures, just as their ships discovered no new continents. To most Europeans the momentary flash of Athenian questioning will seem worth more than all the centuries of Chinese assent.

Yet we must recognize that for thousands of years the Chinese maintained a level of rationality and tolerance that the West might well envy. They had no Index, no Inquisition, no Holy Wars. Superstition has indeed played its part among them; but it has never, as in Europe, been perpetually dominant. It follows from the limitations of Chinese thought that the literature of the country should excel in reflection rather than in speculation. That this is particularly true of its poetry will be gauged from the present volume. In the poems of Po Chü-i no close reasoning or philosophic subtlety will be discovered; but a power of candid reflection and self-analysis which has not been rivalled in the West.

Turning from thought to emotion, the most conspicuous feature of European poetry is its pre-occupation with love. This is apparent not only in actual "love-poems," but in all poetry where the personality of the writer is in any way obtruded. The poet tends to exhibit himself in a *romantic* light; in fact, to recommend himself as a lover.

The Chinese poet has a tendency different but analogous. He recommends himself not as a lover, but as a friend. He poses as a person of infinite leisure (which is what we should most like our friends to possess) and free from worldly ambitions (which constitute the greatest bars to friendship). He would have us think of him as a boon companion, a great drinker of wine, who will not disgrace a social gathering by quitting it sober.

To the European poet the relation between man and woman is a thing of supreme importance and mystery. To the Chinese, it is something commonplace, obvious—a need of the body, not a satisfaction of the *emotions*. These he reserves entirely for friendship.

Accordingly we find that while our poets tend to lay stress on physical courage and other qualities which normal women admire, Po Chü-i is not ashamed to write such a poem as "Alarm at entering the Gorges." Our poets imagine themselves very much as Art has portrayed them—bare-headed and wild-eyed, with shirts unbuttoned at the neck as though they feared that a seizure of emotion might at any minute suffocate them. The Chinese poet introduces himself as a timid recluse, "Reading the Book of Changes at the Northern Window," playing chess with a Taoist priest, or practising caligraphy with an occasional visitor. If "With a Portrait of the Author" had been the rule in the Chinese book-market, it is in such occupations as these that he would be shown; a neat and tranquil figure compared with our lurid frontispieces.

It has been the habit of Europe to idealize love at the expense of friendship and so to place too heavy a burden on the relation of man and woman. The Chinese erred in the opposite direction, regarding their wives and concubines simply as instruments of procreation. For sympathy and intellectual companionship they looked only to their friends. But these friends were bound by no such tie as held women to their masters; sooner or later they drifted away to frontier campaigns, remote governorships, or country retirement. It would not be an exaggeration to say that half the poems in the Chinese language are poems of parting or separation.

Readers of these translations may imagine that the culture represented by Po Chü-i extended over the whole vast confines of China. This would, I think, be an error. Culture is essentially a metropolitan product. Chü-i was as much *dépaysé* at a provincial town as Charles Lamb would have been at Botany Bay. But the system of Chinese bureaucracy tended constantly to break up the literary coteries which formed at the capitals, and to drive the members out of the little corner of Shensi and Honan which to them was "home."

It was chiefly economic necessity which forced the poets of China into the meshes of bureaucracy—backed by the Confucian insistence on public service. To such as were landowners there remained the alternative of agricultural life, arduous and isolated.

The poet, then, usually passed through three stages of existence. In the first we find him with his friends at the capital, drinking, writing, and discussing: burdened by his office probably about as much as Pepys was burdened by his duties at the Admiralty. Next, having failed to curry favour with the Court, he is exiled to some provincial post, perhaps a thousand miles from anyone he cares to talk to. Finally, having scraped together enough money to buy husbands for his daughters, he retires to a small estate, collecting round him the remnants of those with whom he had shared the "feasts and frolics of old days."

I have spoken hitherto only of poets. But the poetess occupies a place of considerable importance in the first four centuries of our era, though the classical period (T'ang and Sung) produced no great woman writer. Her theme varies little; she is almost always a "rejected wife," cast adrift by her lord or sent back to her home. Probably her father would be unable to buy her another husband and there was no place for unmarried women in the Chinese social system. The moment, then, which produced such poems was one of supreme tragedy in a woman's life.

Love-poetry addressed by a man to a woman ceases after

the Han dynasty; but a conventional type of love-poem, in which the poet (of either sex) speaks in the person of a deserted wife or concubine, continues to be popular. The theme appears to be almost an obsession with the T'ang and Sung poets. In a vague way, such poems were felt to be allegorical. Just as in the Confucian interpretation of the love-poems in the Odes (see below) the woman typifies the Minister, and the lover the Prince, so in those classical poems the poet in a veiled way laments the thwarting of his own public ambitions. Such tortuous expression of emotion did not lead to good poetry.

The "figures of speech," devices such as metaphor, simile, and play on words, are used by the Chinese with much more restraint than by us. "Metaphorical epithets" are occasionally to be met with; waves, for example, might perhaps be called "angry." But in general the adjective does not bear the heavy burden which our poets have laid upon it. The Chinese would call the sky "blue," "gray," or "cloudy," according to circumstances; but never "triumphant" or "terror-scourged."

The long Homeric simile, introduced for its own sake or to vary the monotony of narrative, is unknown to Chinese poetry. Shorter similes are sometimes found, as when the half-Chinese poet Altun compares the sky over the Mongolian steppe with the "walls of a tent"; but nothing could be found analogous to Mr. T. S. Eliot's comparison of the sky to a "patient etherized on a table." Except in popular poetry, puns are rare; but there are several characters which, owing to the wideness of their import, are used in a way almost equivalent to play on words.

Classical allusion, always the vice of Chinese poetry, finally destroyed it altogether. In the later periods (from the fourteenth century onwards) the use of elegant synonyms also prevailed. I have before me a "gradus" of the kind which the later poet used as an aid to composition. The moon should be called the "Silver Dish," "Frozen Wheel," or "Golden Ring." Allusions may in this connection be made to Yü Liang, who rode to heaven on the crescent moon; to the hermit T'ang, who controlled the genius of the New Moon, and kept him in his house as a candle—or to any other of some thirty stories which are given. The sun may be called "The Lantern-Dragon," the "Crow in Flight," the "White Colt," etc.

Such were the artificialities of later Chinese poetry.

TECHNIQUE

Certain elements are found, but in varying degree, in all human speech. It is difficult to conceive of a language in which rhyme, stress-accent, and tone-accent would not to some extent occur. In all languages some vowel-sounds are shorter than others and, in certain cases, two consecutive words begin with the same sound. Other such characteristics could be enumerated, but for the purposes of poetry it is these elements which man has principally exploited.

English poetry has used chiefly rhyme, stress, and alliteration. It is doubtful if tone has ever played a part; a conscious use has sporadically been made of quantity. Poetry naturally utilizes the most marked and definite characteristics of the language in which it is written. Such characteristics are used consciously by the poet; but less important elements also play their part, often only in a negative way. Thus the Japanese actually avoid rhyme;

the Greeks did not exploit it, but seem to have tolerated it when it occurred accidentally.

The expedients consciously used by the Chinese before the sixth century were rhyme and length of line. A third element, inherent in the language, was not exploited before that date, but must always have been a factor in instinctive considerations of euphony. This element was "tone."

Chinese prosody distinguishes between two tones, a "flat" and a "deflected." In the first the syllable is enunciated in a level manner: the voice neither rises nor sinks. In the second, it (1) rises, (2) sinks, (3) is abruptly arrested. These varieties make up the Four Tones of Classical Chinese.[1]

The "deflected" tones are distinctly more emphatic, and so have a faint analogy to our stressed syllables. They are also, in an even more remote way, analogous to the long vowels of Latin prosody. A line ending with a "level" has consequently to some extent the effect of a "feminine ending." Certain causes, which I need not specify here, led to an increasing importance of "tone" in the Chinese language from the fifth century onwards. It was natural that this change should be reflected in Chinese prosody. A certain Shēn Yo (A.D. 441-513) first propounded the laws of tone-succession in poetry. From that time till the eighth century the *Lü-shih* or "strictly regulated poem" gradually evolved. But poets continued (and continue till to-day), side by side with their *lü-shih*, to write in the old metre which disregards tone, calling such poems *Ku shih*, "old poems." Previous European statements about Chinese

[1] Not to be confused with the Four Tones of the Mandarin dialect, in which the old names are used to describe quite different enunciations.

prosody should be accepted with great caution. Writers have attempted to define the *lü-shih* with far too great precision.

The Chinese themselves are apt to forget that T'ang poets seldom obeyed the laws designed in later school-books as essential to classical poetry; or, if they notice that a verse by Li Po does not conform, they stigmatize it as "irregular and not to be imitated."

The reader will infer that the distinction between "old poems" and irregular *lü-shih* is often arbitrary. This is certainly the case; I have found the same poem classified differently in different native books. But it is possible to enumerate certain characteristics which distinguish the two kinds of verse. I will attempt to do so; but not till I have discussed *rhyme*, the other main element in Chinese prosody. It would be equally difficult to define accurately the difference between the couplets of Pope and those of William Morris. But it would not be impossible, by pointing out certain qualities of each, to enable a reader to distinguish between the two styles.

Rhyme.—Most Chinese syllables ended with a vowel or nasal sound. The Chinese rhyme was in reality a vowel assonance. Words in different consonants rhymed so long as the vowel-sound was exactly the same. Thus *ywet*, "moon," rhymed with *sek*, "beauty." During the classical period these consonant endings were gradually weakening, and to-day, except in the south, they are wholly lost. It is possible that from very early times final consonants were lightly pronounced.

The rhymes used in *lü-shih* were standardized in the eighth century, and some of them were no longer rhymes to the ear in the Mandarin dialect. To be counted as a rhyme, two words must have exactly the same vowel-sound. Some of the distinctions then made are no longer audible to-day; the sub-divisions therefore seem arbitrary. Absolute homophony is also counted as rhyme, as in French. It is as though we should make *made* rhyme with *maid.*

I will now attempt to distinguish between *Ku-shih* (old style) and *Lü-shih* (new style).

Ku-shih (*Old Style*).

(*a*) According to the investigations of Chu Hua, an eighteenth century critic, only thirty-four rhymes were used. They were, indeed, assonances of the roughest kind.

(*b*) "Deflected" words are used for rhyming as freely as "flat" words.

(*c*) Tone-arrangement. The tones were disregarded. (Lines can be found in pre-T'ang poems in which five deflected tones occur in succession, an arrangement which would have been painful to the ear of a T'ang writer and would probably have been avoided by classical poets even when using the old style.)

Lü-shih (*New Style*).

(*a*) The rhymes used are the "106" of modern dictionaries (not those of the Odes, as Giles states). Rhymes in the flat tone are preferred. In a quatrain the lines which do not rhyme must end on the opposite tone to that of the rhyme. This law is absolute in *Lü-shih* and a tendency in this direction is found even in *Ku-shih.*

(*b*) There is a tendency to antithetical arrangement of tones in the two lines of a couplet, especially in the last part of the lines.

(*c*) A tendency for the tones to go in *pairs*, *e.g.* (A = flat, B = deflected): AA BBA or ABB AA, rather than in *threes.*

Three like tones only come together when divided by a "cesura," *e.g.*, the line BB/AAA would be avoided, but not the line BBAA/ABB.

(*d*) Verbal parallelism in the couplet, *e.g.*:

After long illness one first realizes that seeking medicines is a mistake;
In one's decaying years one begins to repent that one's study of books was deferred.

This device, used with some discretion in T'ang, becomes an irritating trick in the hands of the Sung poets.

THE RISE AND PROGRESS OF CHINESE POETRY

The Odes.—From the songs current in his day Confucius (551-479 B.C.) chose about three hundred which he regarded as suitable texts for his ethical and social teaching. Many of them are eulogies of good rulers or criticisms of bad ones. Out of the three hundred and five still extant only about thirty are likely to interest the modern reader. Of these half deal with war and half with love. Many translations exist, the best being those of Legge in English and of Couvreur in French. There is still room for an English translation displaying more sensitivity to word-rhythm than that of Legge. It should not, I think, include more than fifty poems. But the Odes are essentially *lyric* poetry, and their beauty lies in effects which cannot be reproduced in English. For that reason I have excluded them from this book; nor shall I discuss them further here, for full information will be found in the works of Legge or Couvreur.

Elegies of the Land of Ch'u.—We come next to Ch'ü Yüan (third century B.C.) whose famous poem "Li Sao," or "Falling into Trouble," has also been translated by Legge. It deals, under a love-allegory, with the relation between the writer and his king. In this poem, sex and politics are curiously interwoven, as we need not doubt they were in Chü Yüan's own mind. He affords a striking example of the way in which abnormal mentality imposes itself. We find his followers unsuccessfully attempting to use the same imagery and rhapsodical verbiage, not realizing that these were, as De Goncourt would say, the product of their master's *propre névrosité.*

"The Battle," his one thoroughly intelligible poem, has hitherto been only very imperfectly translated. A literal version will be found on p. 23.

His nephew Sung Yü was no servile imitator. In addition to "elegies" in the style of the Li Sao, he was the author of many "Fu" or descriptive prose-poems, unrhymed but more or less metrical.

The Han Dynasty.—Most of the Han poems in this book were intended to be sung. Many of them are from the official song-book of the dynasty and are known as Yo Fu or Music Bureau poems, as distinct from *shih*, which were recited. Ch'in Chia's poem and his wife's reply (p. 54) are both *shih*; but all the rest might, I think, be counted as songs.

The Han dynasty is rich in Fu (descriptions), but none of them could be adequately translated. They are written in an elaborate and florid style which recalls Apuleius or Lyly.

The Chin Dynasty.

(1) *Popular Songs* (Songs of Wu). The popular songs referred to the Wu (Soochow) district and attributed to the

fourth century may many of them have been current at a much earlier date. They are slight in content and deal with only one topic. They may, in fact, be called "Love-epigrams." They find a close parallel in the *coplas* of Spain, *cf.*:

El candil se està apagando,
La alcuza no tiene aceite—
No te digo que te vayas, . . .
No te digo que te quedes.

The brazier is going out,
The lamp has no more oil—
I do not tell you to go, . . .
I do not tell you to stay.

A Han song, which I will translate quite literally, seems to be the forerunner of the Wu songs.

On two sides of river, wedding made:
Time comes; no boat.
Lusting heart loses hope
Not seeing what-it-desires.

(2) *The Taoists.*—Confucius inculcated the duty of public service. Those to whom this duty was repulsive found support in Taoism, a system which denied this obligation. The third and fourth centuries A.D. witnessed a great reaction against state service. It occurred to the intellectuals of China that they would be happier growing vegetables in their gardens than place-hunting at Nanking. They embraced the theory that "by bringing himself into harmony with Nature" man can escape every evil. Thus Tao (Nature's Way) corresponds to the Nirvana of Buddhism, and the God of Christian mysticism.

They reduced to the simplest standard their houses, apparel, and food; and discarded the load of book-learning which Confucianism imposed on its adherents.

The greatest of these recluses was T'ao Ch'ien (A.D. 365-427), twelve of whose poems will be found on p. 71, *seq.* Something of his philosophy may be gathered from the poem "Substance, Shadow, and Spirit" (p. 73), his own views being voiced by the last speaker. He was not an original thinker, but a great poet who reflects in an interesting way the outlook of his time.

Liang and Minor Dynasties.—This period is known as that of the "Northern and Southern Courts." The north of China was in the hands of the Tungusie Tartars, who founded the Northern Wei dynasty—a name particularly familiar, since it is the habit of European collectors to attribute to this dynasty any sculpture which they believe to be earlier than T'ang. Little poetry was produced in the conquered provinces; the Tartar emperors, though they patronized Buddhist art, were incapable of promoting literature. But at Nanking a series of emperors ruled, most of whom distinguished themselves either in painting or poetry. The Chinese have always (and rightly) despised the literature of this period, which is "all flowers and moonlight." A few individual writers, such as Pao Chao, stand out as exceptions. The Emperor Yüan-ti—who hacked his way to the throne by murdering all other claimants, including his own brother—is typical of the period both as a man and as a poet. A specimen of his sentimental poetry will be found on p. 90. When at last forced to abdicate, he heaped together 200,000 books and pictures; and, setting fire to them, exclaimed: "The culture of the Liang dynasty perishes with me."

T'ang.—I have already described the technical developments of poetry during this dynasty. Form was at this time valued far above content. "Poetry," says a critic, "should draw its materials from the Han and Wei dynasties." With the exception of a few reformers, writers contented themselves with clothing old themes in new forms. The extent to which this is true can of course only be realized by one thoroughly familiar with the earlier poetry.

In the main, T'ang confines itself to a narrow range of stock subjects. The *mise-en-scène* is borrowed from earlier times. If a battle-poem be written, it deals with the campaigns of the Han dynasty, not with contemporary events. The "deserted concubines" of conventional love-poetry are those of the Han Court. Innumerable poems record "Reflections on Visiting a Ruin," or on "The Site of an Old City," etc. The details are ingeniously varied, but the sentiments are in each case identical. Another feature is the excessive use of historical allusions. This is usually not apparent in rhymed translations, which evade such references by the substitution of generalities. Poetry became the medium not for the expression of a poet's emotions, but for the display of his classical attainments. The great Li Po is no exception to this rule. Often where his translators would make us suppose he is expressing a fancy of his own, he is in reality skilfully utilizing some poem by T'ao Ch'ien or Hsieh Ti'ao. It is for his versification that he is admired, and with justice. He represents a reaction against the formal prosody of his immediate predecessors. It was in the irregular song-metres of his *ku-shih* that he excelled. In such poems as the "Ssech'uan Road," with its wild profusion of long and short lines, its cataract of exotic verbiage, he aimed at something nearer akin to music than to poetry. Tu Fu, his contemporary, occasionally abandoned the cult of "abstract form." Both poets lived through the most tragic period of Chinese history. In 755 the Emperor's Turkic favourite, An Lu-shan, revolted against his master. A civil war followed, in which China lost thirty million men. The dynasty was permanently enfeebled and the Empire greatly curtailed by foreign incursions. So ended the "Golden Age" of Ming Huang. Tu Fu, stirred by the horror of massacres and conscriptions, wrote a series of poems in the old style, which Po Chü-i singles out for praise. One of them, "The Press-gang," is familiar in Giles's translation. Li Po, meanwhile, was writing complimentary poems on the Emperor's "Tour in the West"—a journey which was in reality a precipitate flight from his enemies.

Sung.—In regard to content the Sung poets show even less originality than their predecessors. Their whole energy was devoted towards inventing formal restrictions. The "tz'ŭ" developed, a species of song in lines of irregular length, written in strophes, each of which must conform to a strict pattern of tones and rhymes. The content of the "tz'ŭ" is generally wholly conventional. Very few have been translated; and it is obvious that they are unsuitable for translation, since their whole merit lies in metrical dexterity. Examples by the poetess Li I-an will be found in the second edition of Judith Gautier's "Livre de Jade." The poetry of Su Tung-p'o, the foremost writer of the period, is in its matter almost wholly a patchwork of earlier poems. It is for the musical qualities of his verse that he is valued by his countrymen. He hardly wrote a poem which does not contain a phrase (sometimes a whole line) borrowed from Po Chü-i, for whom in his critical writings he expresses boundless admiration.

A word must be said of the Fu (descriptive prose-poems) of this time. They resemble the *vers libres* of modern France, using rhyme occasionally (like Georges Duhamel) as a means of "sonner, rouler, quand il faut faire donner les cuivres et la batterie." Of this nature is the magnificent "Autumn Dirge" (Giles, "Chinese Lit.," p. 215) by Ou-yang Hsiu, whose lyric poetry is of small interest. The subsequent periods need not much concern us. In the eighteenth century the garrulous Yüan Mei wrote his "Anecdotes of Poetry-making"—a book which, while one of the most charming in the language, probably contains more bad poetry (chiefly that of his friends) than any in the world. His own poems are modelled on Po Chü-i and Su Tung-p'o.

This introduction is intended for the general reader. I have therefore stated my views simply and categorically, and without entering into controversies which are of interest only to a few specialists.

As an account of the development of Chinese poetry these notes are necessarily incomplete, but it is hoped that they answer some of those questions which a reader would be most likely to ask.

THE METHOD OF TRANSLATION

IT is commonly asserted that poetry, when literally translated, ceases to be poetry. This is often true, and I have for that reason not attempted to translate many poems which in the original have pleased me quite as much as those I have selected. But I present the ones I have chosen in the belief that they still retain the essential characteristics of poetry.

I have aimed at literal translation, not paraphrase. It may be perfectly legitimate for a poet to borrow foreign themes or material, but this should not be called translation.

Above all, considering imagery to be the soul of poetry, I have avoided either adding images of my own or suppressing those of the original.

Any literal translation of Chinese poetry is bound to be to some extent rhythmical, for the rhythm of the original obtrudes itself. Translating literally, without thinking about the metre of the version, one finds that about two lines out of three have a very definite swing similar to that of the Chinese lines. The remaining lines are just too short or too long, a circumstance very irritating to the reader, whose ear expects the rhythm to continue. I have therefore tried to produce regular rhythmic effects similar to those of the original. Each character in the Chinese is represented by a stress in the English; but between the stresses unstressed syllables are of course interposed. In a few instances where the English insisted on being shorter than the Chinese, I have preferred to vary the metre of my version, rather than pad out the line with unnecessary verbiage.

I have not used rhyme because it is impossible to produce in English rhyme-effects at all similar to those of the original, where the same rhyme sometimes runs through a whole poem. Also, because the restrictions of rhyme necessarily injure either the vigour of one's language or the literalness of one's version. I do not, at any rate, know of any example to the contrary. What is generally known as "blank verse" is the worst medium for translating Chinese poetry, because the essence of blank verse is that it varies the position of its pauses, whereas in Chinese the stop always comes at the end of the couplet.

BIOGRAPHICAL NOTES

1. H. A. Giles, "Chinese Poetry in English Verse." 1896. 212 pp. Combines rhyme and literalness with wonderful dexterity.

2. Hervey St. Denys, "Poésies des Thang." 1862. 301 pp. The choice of poems would have been very different if the author had selected from the whole range of T'ang poetry, instead of contenting himself, except in the case of Li Po and Tu Fu, with making extracts from two late anthologies. This book, the work of a great scholar, is reliable—except in its information about Chinese prosody.

3. Judith Gautier, "Le Livre de Jade." 1867 and 1908. It has been difficult to compare these renderings with the original, for proper names are throughout distorted or interchanged. For example, part of a poem by Po Chü-i *about* Yang T'ai-chēn is here given as a complete poem and ascribed to "Yan-Ta-Tchen" as author. The poet Han Yü figures as Heu-Yu; T'ao Han as Sao Nan, etc. Such mistakes are evidently due to faulty decipherment of someone else's writing. Nevertheless, the book is far more readable than that of St. Denys, and shows a wider acquaintance with Chinese poetry on the part of whoever chose the poems. Most of the credit for this selection must certainly be given to Ting Tun-ling, the *literatus* whom Théophile Gautier befriended. But the credit for the beauty of these often erroneous renderings must go to Mademoiselle Gautier herself.

4. Anna von Bernhardi, in "Mitteil d. Seminar f. Orient. Sprachen," 1912, 1915, and 1916. Two articles of T'ao Ch'ien and one on Li Po. All valuable, though not free from mistakes.

5. Zottoli, "Cursus Litteraturae Sinicae." 1886. Chinese text with Latin translation. Vol. V deals with poetry. None of the poems is earlier than T'ang. The Latin is seldom intelligible without reference to the Chinese. Translators have obviously used Zottoli as a text. Out of eighteen Sung poems in Giles's book, sixteen will be found in Zottoli.

6. A. Pfizmaier, two articles (1886 and 1887) on Po Chü-i in "Denkschr. d. Kais. Ak. in Wien." So full of mistakes as to be of

very little value, except in so far as they served to call the attention of the European reader to this poet.

7. L. Woitsch, "Aus den Gedichten Po Chü-i's." 1908. 76 pp. A prose rendering with Chinese text of about forty poems, not very well selected. The translations, though inaccurate, are a great advance on Pfizmaier.

8. E. von Zachs, "Lexicographische Beiträge." Vols. ii and iv. Re-translation of two poems previously mistranslated by Pfizmaier.

9. S. Imbault-Huart, "La Poésie Chinoise du 14 au 19 siècle." 1886. 93 pp.

10. S. Imbault-Huart, "Un Poète Chinois du 18 Siècle." (Yüan Mei.) Journ. of China Branch, Royal As. Soc., N.S., vol. xix, part 2, 42 pp.

11. S. Imbault-Huart, "Poésies Modernes." 1892. 46 pp.

12. A. Forke, "Blüthen Chinesischer Dichtung." 1899. Rhymed versions of Li Po and pre-T'ang poems.

A fuller bibliography will be found in Cordier's "Bibliotheca Sinica."

CHAPTER I

BATTLE

By Ch'ü Yüan (332-295 B.C.), author of the famous poem "Li Sao, or "Falling into Trouble." Finding that he could not influence the conduct of his prince, he drowned himself in the river Mi-lo. The modern Dragon Boat Festival is supposed to be in his honour.

"WE grasp our battle-spears: we don our breast-plates of hide.
The axles of our chariots touch: our short swords meet.
Standards obscure the sun: the foe roll up like clouds.
Arrows fall thick: the warriors press forward.
They menace our ranks: they break our line.
The left-hand trace-horse is dead: the one on the right is smitten.
The fallen horses block our wheels: they impede the yoke-horses!"

They grasp their jade drum-sticks: they beat the sounding drums.
Heaven decrees their fall: the dread Powers are angry.

The warriors are all dead: they lie on the moor-field.
They issued but shall not enter: they went but shall not return.
The plains are flat and wide: the way home is long.

Their swords lie beside them: their black bows, in their hand.
Though their limbs were torn, their hearts could not be repressed.
They were more than brave: they were inspired with the spirit of "Wu."[1]
Steadfast to the end, they could not be daunted.
Their bodies were stricken, but their souls have taken Immortality—
Captains among the ghosts, heroes among the dead.

THE MAN-WIND AND THE WOMAN-WIND

A "fu," or prose-poem, by Sung Yü (fourth century B.C.), nephew of Ch'ü Yüan.

HSIANG, king of Ch'u, was feasting in the Orchid-tower Palace, with Sung Yü and Ching Ch'ai to wait upon him. A gust of wind blew in and the king bared his breast to meet it, saying: "How pleasant a thing is this wind which I share with the common people." Sung Yü answered: "This is the Great King's wind. The common people cannot share it." The king said: "Wind is a spirit of Heaven and Earth. It comes wide spread and does not choose between noble and base or between high and low. How can you say 'This is the king's wind'?" Sung answered: "I have heard it taught that in the crooked lemon-tree birds make their nests and to empty spaces winds fly. But the wind-spirit that comes to different things is not the same." The king said: "Where is the wind born?" and

[1] *I.e.*, military genius.

Sung answered: "The wind is born in the ground. It rises in the extremities of the green p'ing-flower. It pours into the river-valleys and rages at the mouth of the pass. It follows the rolling flanks of Mount T'ai and dances beneath the pine-trees and cypresses. In gusty bouts it whirls. It rushes in fiery anger. It rumbles low with a noise like thunder, tearing down rocks and trees, smiting forests and grasses.

"But at last abating, it spreads abroad, seeks empty places and crosses the threshold of rooms. And so growing gentler and clearer, it changes and is dispersed and dies.

"It is this cool clear Man-Wind that, freeing itself, falls and rises till it climbs the high walls of the Castle and enters the gardens of the Inner Palace. It bends the flowers and leaves with its breath. It wanders among the osmanthus and pepper-trees. It lingers over the fretted face of the pond, to steal the soul of the hibiscus. It touches the willow leaves and scatters the fragrant herbs. Then it pauses in the courtyard and turning to the North goes up to the Jade Hall, shakes the hanging curtains and lightly passes into the inner room.

"And so it becomes the Great King's wind.

"Now such a wind is fresh and sweet to breathe and its gentle murmuring cures the diseases of men, blows away the stupor of wine, sharpens sight and hearing and refreshes the body. This is what is called the Great King's wind."

The king said: "You have well described it. Now tell me of the common people's wind." Sung said: "The common people's wind rises from narrow lanes and streets, carrying clouds of dust. Rushing to empty spaces it attacks the gateway, scatters the dust-heap, sends the cinders flying, pokes among foul and rotting things, till at last it enters the tiled windows and reaches the rooms of the cottage. Now this wind is heavy and turgid, oppressing man's heart. It brings fever to his body, ulcers to his lips and dimness to his eyes. It shakes him with coughing; it kills him before his time.

"Such is the Woman-wind of the common people."

The following is a sample of Sung Yü's prose:

MASTER TĒNG-T'U

By Sung Yü (third century B.C.)

ONE day when the Chamberlain, master Tēng-t'u, was in attendance at the Palace he warned the King against Sung Yü, saying: "Yü is a man of handsome features and calm bearing and his tongue is prompt with subtle sentences. Moreover, his character is licentious. I would submit that your Majesty is ill-advised in allowing him to follow you into the Queen's apartments." The King repeated Tēng-t'u's words to Sung Yü. Yü replied: "My beauty of face and calmness of bearing were given me by Heaven. Subtlety of speech I learnt from my teachers. As for my character, I deny that it is licentious." The King said: "Can you substantiate your statement that you are not licentious? If you cannot, you must leave the Court." Sung Yü said: "Of all the women in the world, the most beautiful are the women of the land of Ch'u. And in all the land of Ch'u there are none like the women of my own village. And in my village there are none that can be compared with the girl next door

"The girl next door would be too tall if an inch were added to her height, and too short if an inch were taken away. Another grain of powder would make her too pale; another touch of rouge would make her too red. Her eyebrows are like the plumage of the kingfisher, her flesh is like snow. Her waist is like a roll of new silk, her teeth are like little shells. A single of her smiles would perturb the whole city of Yang and derange the suburb of Hsia-ts'ai.[1] For three years this lady has been climbing the garden wall and peeping at me, yet I have never succumbed.

"How different is the behaviour of master Tēng-t'u! His wife has a wooly head and misshapen ears; projecting teeth irregularly set; a crook in her back and a halt in her gait. Moreover, she has running sores in front and behind.

"Yet Tēng-t'u fell in love with her and caused her to bear him five children.

"I would have your Majesty consider which of us is the debauchee."

Sung Yü was not dismissed from court.

THE ORPHAN

Anon. (first century B.C.)

To be an orphan,
To be fated to be an orphan,
How bitter is this lot!
When my father and mother were alive
I used to ride in a carriage
With four fine horses.

[1] Fashionable quarters in the capital of Ch'u state.

But when they both died,
My brother and sister-in-law
Sent me out to be a merchant.
In the south I travelled to the "Nine Rivers"
And in the east as far as Ch'i and Lu.
At the end of the year when I came home
I dared not tell them what I had suffered—
Of the lice and vermin in my head,
Of the dust in my face and eyes.
My brother told me to get ready the dinner,
My sister-in-law told me to see after the horses.
I was always going up into the hall
And running down again to the parlour.
My tears fell like rain.

In the morning they sent me to draw water,
I didn't get back till night-fall.
My hands were all sore
And I had no shoes.
I walked the cold earth
Treading on thorns and brambles.
As I stopped to pull out the thorns,
How bitter my heart was!
My tears fell and fell
And I went on sobbing and sobbing.
In winter I have no great-coat;
Nor in summer, thin clothes.
It is no pleasure to be alive.
I had rather quickly leave the earth
And go beneath the Yellow Springs.[1]

[1] Hades.

The April winds blow
And the grass is growing green.
In the third month—silkworms and mulberries,
In the sixth month—the melon-harvest.
I went out with the melon-cart
And just as I was coming home
The melon-cart turned over.
The people who came to help me were few,
But the people who eat the melons were many,
All they left me was the stalks—
To take home as fast as I could.
My brother and sister-in-law were harsh,
They asked me all sorts of awful questions.
Why does everyone in the village hate me?
I want to write a letter and send it
To my mother and father under the earth,
And tell them I can't go on any longer
Living with my brother and sister-in-law.

THE SICK WIFE

SHE had been ill for years and years;
She sent for me to say something.
She couldn't say what she wanted
Because of the tears that kept coming of themselves.
"I have burdened you with orphan children,
With orphan children two or three.
Don't let our children go hungry or cold;
If they do wrong, don't slap or beat them.
When you take out the baby, rock it in your arms.
Don't forget to do that."

Last she said,
"When I carried them in my arms they had no clothes
And now their jackets have no linings." [*She dies.*

I shut the doors and barred the windows
And left the motherless children.
When I got to the market and met my friends, I wept.
I sat down and could not go with them.
I asked them to buy some cakes for my children.
In the presence of my friends I sobbed and cried.
I tried not to grieve, but sorrow would not cease.
I felt in my pocket and gave my friends some money.
When I got home I found my children
Calling to be taken into their mother's arms.
I walked up and down in the empty room
This way and that a long while.
Then I went away from it and said to myself
"I will forget and never speak of her again."

COCK-CROW SONG

Anon. (first century B.C.)

IN the eastern quarter dawn breaks, the stars flicker pale.
The morning cock at Ju-nan mounts the wall and crows.
The songs are over, the clock[1] run down, but still the feast is set.
The moon grows dim and the stars are few; morning has come to the world.

[1] A water-clock.

At a thousand gates and ten thousand doors the fish-shaped keys turn;
Round the Palace and up by the Castle, the crows and magpies are flying.

THE GOLDEN PALACE

Anon. (first century B.C.)

We go to the Golden Palace:
We set out the jade cups.
We summon the honoured guests
To enter at the Golden Gate.
They enter at the Golden Gate
And go to the Golden Hall.
In the Eastern Kitchen the meat is sliced and ready—
Roast beef and boiled pork and mutton.
The Master of the Feast hands round the wine.
The harp-players sound their clear chords.

The cups are pushed aside and we face each other at chess:
The rival pawns are marshalled rank against rank.
The fire glows and the smoke puffs and curls;
From the incense-burner rises a delicate fragrance.
The clear wine has made our cheeks red;
Round the table joy and peace prevail.
May those who shared in this day's delight
Through countless autumns enjoy like felicity.

"OLD POEM"

At fifteen I went with the army,
At fourscore I came home.
On the way I met a man from the village,
I asked him who there was at home.
"That over there is your house,
All covered over with trees and bushes."
Rabbits had run in at the dog-hole,
Pheasants flew down from the beams of the roof.
In the courtyard was growing some wild grain;
And by the well, some wild mallows.
I'll boil the grain and make porridge,
I'll pluck the mallows and make soup.
Soup and porridge are both cooked,
But there is no one to eat them with.
I went out and looked towards the east,
While tears fell and wetted my clothes.

MEETING IN THE ROAD

In a narrow road where there was not room to pass
My carriage met the carriage of a young man.
And while his axle was touching my axle
In the narrow road I asked him where he lived.
"The place where I live is easy enough to find,
Easy to find and difficult to forget.
The gates of my houses are built of yellow gold,
The hall of my house is paved with white jade,

On the hall table flagons of wine are set,
I have summoned to serve me dancers of Han-tan.[1]
In the midst of the courtyard grows a cassia-tree,—
And candles on its branches flaring away in the night.

FIGHTING SOUTH OF THE CASTLE

Anon. (*circa* 124 B.C.)

They fought south of the Castle,
They died north of the wall.
They died in the moors and were not buried.
Their flesh was the food of crows.
"Tell the crows we are not afraid;
We have died in the moors and cannot be buried.
Crows, how can our bodies escape you?"
The waters flowed deep
And the rushes in the pool were dark.
The riders fought and were slain:
Their horses wander neighing.
By the bridge there was a house.[2]
Was it south, was it north?
The harvest was never gathered.
How can we give you your offerings?
You served your Prince faithfully,
Though all in vain.

[1] Capital of the kingdom of Chao, where the people were famous for their beauty.

[2] There is no trace of it left. This passage describes the havoc of war. The harvest has not been gathered: therefore corn-offerings cannot be made to the spirits of the dead.

I think of you, faithful soldiers;
Your service shall not be forgotten.
For in the morning you went out to battle
And at night you did not return.

THE EASTERN GATE

Anon. (first century B.C.).

A poor man determines to go out into the world and make his fortune. His wife tries to detain him.

I WENT out at the eastern gate:
I never thought to return.
But I came back to the gate with my heart full of sorrow.

There was not a peck of rice in the bin:
There was not a coat hanging on the pegs.
So I took my sword and went towards the gate.
My wife and child clutched at my coat and wept:
"Some people want to be rich and grand:
I only want to share my porridge with you.
Above, we have the blue waves of the sky:
Below, the yellow face of this little child."
"Dear wife, I cannot stay.
Soon it will be too late.
When one is growing old
One cannot put things off."

OLD AND NEW

Anon. (first century B.C.)

SHE went up the mountain to pluck wild herbs;
She came down the mountain and met her former husband.
She knelt down and asked her former husband
"What do you find your new wife like?"
"My new wife, although her talk is clever,
Cannot charm me as my old wife could.
In beauty of face there is not much to choose,
But in usefulness they are not at all alike.
My new wife comes in from the road to meet me;
My old wife always came down from her tower.
My new wife is clever at embroidering silk;
My old wife was good at plain sewing.
Of silk embroidery one can do an inch a day;
Of plain sewing, more than five feet.
Putting her silks by the side of your sewing,
I see that the new will not compare with the old."

SOUTH OF THE GREAT SEA

MY love is living
To the south of the Great Sea.
What shall I send to greet him?
Two pearls and a comb of tortoise-shell:
I'll send them to him packed in a box of jade.

They tell me he is not true:
They tell me he dashed my box to the ground,
Dashed it to the ground and burnt it
And scattered its ashes to the wind.
From this day to the ends of time
I must never think of him,
Never again think of him.
The cocks are crowing,
And the dogs are barking—
My brother and his wife will soon know.[1]
The autumn wind is blowing;
The morning wind is sighing.
In a moment the sun will rise in the east
And then *it* too will know.

THE OTHER SIDE OF THE VALLEY

I AM a prisoner in the hands of the enemy,
Enduring the shame of captivity.
My bones stick out and my strength is gone
Through not getting enough to eat.
My brother is a Mandarin
And his horses are fed on maize.
Why can't he spare a little money
To send and ransom me?

[1] *I.e.*, about her engagement being broken off.

OATHS OF FRIENDSHIP

In the country of Yüeh when a man made friends with another they set up an altar of earth and sacrificed upon it a dog and a cock, reciting this oath as they did so:

(1)

If you were riding in a coach
And I were wearing a "li,"[1]
And one day we met in the road,
You would get down and bow.
If you were carrying a "tēng"[2]
And I were riding on a horse,
And one day we met in the road
I would get down for you.

(2)

Shang Ya!
I want to be your friend
For ever and ever without break or decay.
When the hills are all flat
And the rivers are all dry,
When it lightens and thunders in winter,
When it rains and snows in summer,
When Heaven and Earth mingle—
Not till then will I part from you.

[1] A peasant's coat made of straw.
[2] An umbrella under which a cheap-jack sells his wares.

BURIAL SONGS

(1)

"The dew on the garlic-leaf," sung at the burial of kings and princes.

How swiftly it dries,
The dew on the garlic-leaf,
The dew that dries so fast
To-morrow will fall again.
But he whom we carry to the grave
Will never more return.

(2)

"The Graveyard," sung at the burial of common men.

WHAT man's land is the graveyard?
It is the crowded home of ghosts,—
Wise and foolish shoulder to shoulder.
The King of the Dead claims them all;
Man's fate knows no tarrying.

SEVENTEEN OLD POEMS

The following seventeen poems are from a series known as the Nineteen Pieces of Old Poetry. Some have been attributed to Mei Shēng (first century B.C.), and one to Fu I (first century A.D.). They are manifestly not all by the same hand nor of the same date. Internal evidence shows that No. 3 at least was written after the date of Mei Shēng's death. These poems had an enormous influence on all subsequent poetry, and many of the habitual *clichés* of Chinese verse are taken from them. I have omitted two because of their marked inferiority.

(1)

ON and on, always on and on
Away from you, parted by a life-parting.[1]
Going from one another ten thousand "li,"
Each in a different corner of the World.
The way between is difficult and long,
Face to face how shall we meet again?
The Tartar horse prefers the North wind,
The bird from Yüeh nests on the Southern branch.
Since we parted the time is already long,
Daily my clothes hang looser round my waist.
Floating clouds obscure the white sun,
The wandering one has quite forgotten home.
Thinking of you has made me suddenly old,
The months and years swiftly draw to their close.
I'll put you out of my mind and forget for ever
And try with all my might to eat and thrive.[2]

[1] The opposite of a parting by death.

[2] The popular, but erroneous, interpretation of these two lines is:

> "That I'm cast away and rejected I will not repine,
> But only hope with all my heart you're well."

(2)

GREEN, green,
The grass by the river-bank.
Thick, thick,
The willow trees in the garden.
Sad, sad,
The lady in the tower.
White, white,
Sitting at the casement window.
Fair, fair,
Her red-powdered face.
Small, small,
She puts out her pale hand.
Once she was a dancing-house girl,
Now she is a wandering man's wife.
The wandering man went, but did not return.
It is hard alone to keep an empty bed.

(3)

GREEN, green,
The cypress on the mound.
Firm, firm,
The boulder in the stream.
Man's life lived within this world,
Is like the sojourning of a hurried traveller.
A cup of wine together will make us glad,
And a little friendship is no little matter.

Yoking my chariot I urge my stubborn horses.
I wander about in the streets of Wan and Lo.
In Lo Town how fine everything is!

The "Caps and Belts"[1] go seeking each other out.
The great boulevards are intersected by lanes,
Wherein are the town-houses of Royal Dukes.
The two palaces stare at each other from afar,
The twin gates rise a hundred feet.
By prolonging the feast let us keep our hearts gay,
And leave no room for sadness to creep in.

(4)

Of this day's glorious feast and revel
The pleasure and delight are difficult to describe.
Plucking the lute they sent forth lingering sounds,
The new melodies in beauty reached the divine.
Skilful singers intoned the high words,
Those who knew the tune heard the trueness of their singing.
We sat there each with the same desire
And like thoughts by each unexpressed:
"Man in the world lodging for a single life-time
Passes suddenly like dust borne on the wind.
Then let us hurry out with high steps
And be the first to reach the highways and fords:
Rather than stay at home wretched and poor
For long years plunged in sordid grief."

(5)

In the north-west there is a high house,
Its top level with the floating clouds.
Embroidered curtains thinly screen its windows,
Its storied tower is built on three steps.

[1] High officers.

From above there comes a noise of playing and singing,
The tune sounding, oh! how sad!
Who can it be, playing so sad a tune?
Surely it must be Ch'i Liang's[1] wife.
The tranquil "D" follows the wind's rising,
The middle lay lingers indecisive.
To each note, two or three sobs,
Her high will conquered by overwhelming grief.
She does not regret that she is left so sad,
But minds that so few can understand her song.
She wants to become those two wild geese
That with beating wings rise high aloft.

(6)

CROSSING the river I pluck hibiscus-flowers:
In the orchid-swamps are many fragrant herbs.
I gather them, but who shall I send them to?
My love is living in lands far away.
I turn and look towards my own country:
The long road stretches on for ever.
The same heart, yet a different dwelling:
Always fretting, till we are grown old!

(7)

A BRIGHT moon illumines the night-prospect:
The house-cricket chirrups on the eastern wall.
The Handle of the Pole-star points to the Beginning of
Winter.
The host of stars is scattered over the sky.

[1] Who had no father, no husband, and no children.

The white dew wets the moor-grasses,—
With sudden swiftness the times and seasons change.
The autumn cicada sings among the trees,
The swallows, alas, whither are they gone?

Once I had a same-house friend,
He took flight and rose high away.
He did not remember how once we went hand in hand,
But left me like footsteps behind one in the dust.

In the South is the Winnowing-fan and the Pole-star in the North,
And a Herd-boy [1] whose ox has never borne the yoke.
A friend who is not firm as a great rock
Is of no profit and idly bears the name.

(8)

In the courtyard there grows a strange tree,
Its green leaves ooze with a fragrant moisture.
Holding the branch I cut a flower from the tree,
Meaning to send it away to the person I love.
Its sweet smell fills my sleeves and lap.
The road is long, how shall I get it there?
Such a thing is not fine enough to send:
But it may remind him of the time that has past since he left.[2]

[1] Name of a star. The Herd-boy, who is only figuratively speaking a herd-boy, is like the friend who is no real friend.

[2] *I.e.* (supposing he went away in the autumn), remind him that spring has come.

(9)

FAR away twinkles the Herd-boy star;
Brightly shines the Lady of the Han River.
Slender, slender she plies her white fingers.
Click, click go the wheels of her spinning-loom.
At the end of the day she has not finished her task;
Her bitter tears fall like streaming rain.
The Han River runs shallow and clear;
Set between them, how short a space!
But the river water will not let them pass,
Gazing at each other but never able to speak.

(10)

TURNING my chariot I yoke my horses and go.
On and on down the long roads
The autumn winds shake the hundred grasses.
On every side, how desolate and bare!
The things I meet are all new things,
Their strangeness hastens the coming of old age.
Prosperity and decay each have their season.
Success is bitter when it is slow in coming.
Man's life is not metal or stone,
He cannot far prolong the days of his fate.
Suddenly he follows in the way of things that change.
Fame is the only treasure that endures.

(11)

THE Eastern Castle stands tall and high;
Far and wide stretch the towers that guard it.
The whirling wind uprises and shakes the earth;
The autumn grasses grow thick and green.

The four seasons alternate without pause,
The year's end hurries swiftly on.
The Bird of the Morning Wind is stricken with sorrow;
The frail cicada suffers and is hard pressed.
Free and clear, let us loosen the bonds of our hearts.
Why should we go on always restraining and binding?
In Yen and Chao are many fair ladies,
Beautiful people with faces like jade.
Their clothes are made all of silk gauze,
They stand at the door practising tranquil lays.
The echo of their singing, how sad it sounds!
By the pitch of the song one knows the stops have been
tightened.
To ease their minds they arrange their shawls and belts;
Lowering their song, a little while they pause.
"I should like to be those two flying swallows
Who are carrying clay to nest in the eaves of your house."

(12)

I DRIVE my chariot up to the Eastern Gate;
From afar I see the graveyard north of the Wall.
The white aspens how they murmur, murmur;
Pines and cypresses flank the broad paths.
Beneath lie men who died long ago;
Black, black is the long night that holds them.
Deep down beneath the Yellow Springs,
Thousands of years they lie without waking.

In infinite succession light and darkness shift,
And years vanish like the morning dew.
Man's life is like a sojourning,
His longevity lacks the firmness of stone and metal.

For ever it has been that mourners in their turn were
mourned,
Saint and Sage,—all alike are trapped.
Seeking by food to obtain Immortality
Many have been the dupe of strange drugs.
Better far to drink good wine
And clothe our bodies in robes of satin and silk.

(13) Continuation of (12)

The dead are gone and with them we cannot converse.
The living are here and ought to have our love.
Leaving the city-gate I look ahead
And see before me only mounds and tombs.
The old graves are ploughed up into fields,
The pines and cypresses are hewn for timber.
In the white aspens sad winds sing;
Their long murmuring kills my heart with grief.
I want to go home, to ride to my village gate.
I want to go back, but there's no road back.

(14)

The years of a lifetime do not reach a hundred,
Yet they contain a thousand years' sorrow.
When days are short and the dull nights long,
Why not take a lamp and wander forth?
If you want to be happy you must do it now,
There is no waiting till an after-time.
The fool who 's loath to spend the wealth he 's got
Becomes the laughing-stock of after ages.
It is true that Master Wang became immortal,
But how can *we* hope to share his lot?

(15)

COLD, cold the year draws to its end,
The crickets and grasshoppers make a doleful chirping.
The chill wind increases its violence.
My wandering love has no coat to cover him.
He gave his embroidered furs to the Lady of Lo,
But from me his bedfellow he is quite estranged.
Sleeping alone in the depth of the long night
In a dream I thought I saw the light of his face.
My dear one thought of our old joys together,
He came in his chariot and gave me the front reins.
I wanted so to prolong our play and laughter,
To hold his hand and go back with him in his coach.
But, when he had come he would not stay long
Nor stop to go with me to the Inner Chamber.
Truly without the falcon's wings to carry me
How can I rival the flying wind's swiftness?
I go and lean at the gate and think of my grief,
My falling tears wet the double gates.

(16)

AT the beginning of winter a cold spirit comes,
The North Wind blows—chill, chill.
My sorrows being many, I know the length of the nights,
Raising my head I look at the stars in their places.
On the fifteenth day the bright moon is full,
On the twentieth day the "toad and hare" wane.[1]
A stranger came to me from a distant land
And brought me a single scroll with writing on it;

[1] The "toad and hare" correspond to our "man in the moon." The waning of the moon symbolizes the waning of the lover's affection.

At the top of the scroll was written "Do not forget,"
At the bottom was written "Goodbye for Ever."
I put the letter away in the folds of my dress,
For three years the writing did not fade.
How with an undivided heart I loved you
I fear that you will never know or guess.

(17)

THE bright moon, oh, how white it shines,
Shines down on the gauze curtains of my bed.
Racked by sorrow I toss and cannot sleep.
Picking up my clothes, I wander up and down.
My absent love says that he is happy,
But I would rather he said he was coming back.
Out in the courtyard I stand hesitating, alone.
To whom can I tell the sad thoughts I think?
Staring before me I enter my room again;
Falling tears wet my mantle and robe.

THE AUTUMN WIND

By Wu-ti (157-87 B.C.), sixth emperor of the Han dynasty. He came to the throne when he was only sixteen. In this poem he regrets that he is obliged to go on an official journey, leaving his mistress behind in the capital. He is seated in his state barge surrounded by his ministers.

AUTUMN wind rises: white clouds fly.
Grass and trees wither: geese go south.
Orchids all in bloom: chrysanthemums smell sweet.
I think of my lovely lady: I never can forget.

Floating-pagoda boat crosses Fēn River.
Across the mid-stream white waves rise
Flute and drum keep time to sound of the rowers' song;
Amidst revel and feasting, sad thoughts come;
Youth's years how few! Age how sure!

LI FU-JĒN

THE sound of her silk skirt has stopped.
On the marble pavement dust grows.
Her empty room is cold and still.
Fallen leaves are piled against the doors.
Longing for that lovely lady
How can I bring my aching heart to rest?

The above poem was written by Wu-ti when his mistress, Li Fu-jēn, died. Unable to bear his grief, he sent for wizards from all parts of China, hoping that they would be able to put him into communication with her spirit. At last one of them managed to project her shape on to a curtain. The emperor cried:

Is it or isn't it?
I stand and look.
The swish, swish of a silk skirt.
How slow she comes!

SONG OF SNOW-WHITE HEADS

Ssŭ-ma Hsiang-ju was a young poet who had lost his position at court owing to ill-health. One day Cho Wĕn-chün, a rich man's daughter, heard him singing at a feast given by her father. She eloped with him that night, and they set up a wine-shop together. After a time Hsiang-ju became famous as a poet, but his character was marred by love of money. He sold love-poems, which the ladies of the palace sent to the emperor in order to win his favour. Finally, he gave presents to the "ladies of Mo-ling," hoping to secure a concubine. It was this step that induced his mistress, Cho Wĕn-chün, to write the following poem.

OUR love was pure
As the snow on the mountains:
White as a moon
Between the clouds—
They're telling me
Your thoughts are double
That's why I've come
To break it off.
To-day we'll drink
A cup of wine.
To-morrow we'll part
Beside the Canal:
Walking about
Beside the Canal,
Where its branches divide
East and west.
Alas and alas,
And again alas.
So must a girl
Cry when she's married,

If she find not a man
Of single heart,
Who will not leave her
Till her hair is white.

TO HIS WIFE

By General Su Wu (*circa* 100 B.C.)

Since our hair was plaited and we became man and wife
The love between us was never broken by doubt.
So let us be merry this night together,
Feasting and playing while the good time lasts.

I suddenly remember the distance that I must travel;
I spring from bed and look out to see the time.
The stars and planets are all grown dim in the sky;
Long, long is the road; I cannot stay.
I am going on service, away to the battle-ground,
And I do not know when I shall come back.
I hold your hand with only a deep sigh;
Afterwards, tears—in the days when we are parted.
With all your might enjoy the spring flowers,
But do not forget the time of our love and pride.
Know that if I live, I will come back again,
And if I die, we will go on thinking of each other.

LI LING

(Parting from Su Wu)

THE good time will never come back again:
In a moment,—our parting will be over.
Anxiously—we halt at the road-side,
Hesitating—we embrace where the fields begin.
The clouds above are floating across the sky:
Swiftly, swiftly passing: or blending together.
The waves in the wind lose their fixed place
And are rolled away each to a corner of Heaven.
From now onwards—long must be our parting,
So let us stop again for a little while.
I wish I could ride on the wings of the morning wind
And go with you right to your journey's end.

Li Ling and Su Wu were both prisoners in the land of the Huns. After nineteen years Su Wu was released. Li Ling would not go back with him. When invited to do so, he got up and danced, singing:

I came ten thousand leagues
Across sandy deserts
In the service of my Prince,
To break the Hun tribes.
My way was blocked and barred,
My arrows and sword broken.
My armies had faded away,
My reputation had gone.

My old mother is long dead.
Although I want to requite my Prince
How can I return?

LAMENT OF HSI-CHÜN

About the year 110 B.C. a Chinese Princess named Hsi-chün was sent, for political reasons, to be the wife of a central Asian nomad king, K'un Mo, king of the Wu-sun. When she got there, she found her husband old and decrepit. He only saw her once or twice a year, when they drank a cup of wine together. They could not converse, as they had no language in common.

My people have married me
In a far corner of Earth:
Sent me away to a strange land,
To the king of the Wu-sun.
A tent is my house,
Of felt are my walls;
Raw flesh my food
With mare's milk to drink.
Always thinking of my own country,
My heart sad within.
Would I were a yellow stork
And could fly to my old home!

CH'IN CHIA

Ch'in Chia (first century A.D.) was summoned to take up an appointment at the capital at a time when his wife was ill and staying with her parents. He was therefore unable to say goodbye to her, and sent her three poems instead. This is the last of the three.

Solemn, solemn the coachman gets ready to go:
"Chiang, chiang" the harness bells ring.
At break of dawn I must start on my long journey:
At cock-crow I must gird on my belt.

I turn back and look at the empty room:
For a moment I almost think I see you there.
One parting, but ten thousand regrets:
As I take my seat, my heart is unquiet.
What shall I do to tell you all my thoughts?
How can I let you know of all my love?
Precious hairpins make the head to shine
And bright mirrors can reflect beauty.
Fragrant herbs banish evil smells
And the scholar's harp has a clear note.
The man in the Book of Odes [1] who was given a quince
Wanted to pay it back with diamonds and rubies.
When I think of all the things you have done for me,
How ashamed I am to have done so little for you!
Although I know that it is a poor return,
All I can give you is this description of my feelings.

CH'IN CHIA'S WIFE'S REPLY

My poor body is alas unworthy:
I was ill when first you brought me home.
Limp and weary in the house—
Time passed and I got no better.
We could hardly ever see each other:
I could not serve you as I ought.
Then you received the Imperial Mandate:
You were ordered to go far away to the City.
Long, long must be our parting:
I was not destined to tell you my thoughts.

[1] Odes, v, 10.

I stood on tiptoe gazing into the distance,
Interminably gazing at the road that had taken you.
With thoughts of you my mind is obsessed:
In my dreams I see the light of your face.
Now you are started on your long journey
Each day brings you further from me.
Oh that I had a bird's wings
And high flying could follow you.
Long I sob and long I cry:
The tears fall down and wet my skirt.

SONG

By Sung Tzŭ-hou (second century A.D.)

On the Eastern Way at the city of Lo-yang
At the edge of the road peach-trees and plum-trees grow;
On the two sides,—flower matched by flower;
Across the road,—leaf touching leaf.

A spring wind rises from the north-east;
Flowers and leaves gently nod and sway.
Up the road somebody's daughter comes
Carrying a basket, to gather silkworms' food.
(*She sees the fruit trees in blossom and, forgetting about her silkworms, begins to pluck the branches.*)
With her slender hand she breaks a branch from the tree;
The flowers fall, tossed and scattered in the wind.

The tree says:

" Lovely lady, I never did you harm;
Why should you hate me and do me injury? "

The lady answers:

" At high autumn in the eighth and ninth moons
When the white dew changes to hoar-frost,
At the year's end the wind would have lashed your boughs,
Your sweet fragrance could not have lasted long.
Though in the autumn your leaves patter to the ground,
When spring comes, your gay bloom returns.
But in men's lives when their bright youth is spent
Joy and love never come back again.

CHAPTER II

SATIRE ON PAYING CALLS IN AUGUST

By Ch'ēng Hsiao (*circa* A.D. 250)

WHEN I was young, throughout the hot season
There were no carriages driving about the roads.
People shut their doors and lay down in the cool:
Or if they went out, it was not to pay calls.
Nowadays—ill-bred, ignorant fellows,
When they feel the heat, make for a friend's house.
The unfortunate host, when he hears someone coming
Scowls and frowns, but can think of no escape.
"There 's nothing for it but to rise and go to the door,"
And in his comfortable seat he groans and sighs.

The conversation does not end quickly:
Prattling and babbling, what a lot he says!
Only when one is almost dead with fatigue
He asks at last if one isn't finding him tiring.
(One's arm is almost in half with continual fanning:
The sweat is pouring down one's neck in streams.)
Do not say that this is a small matter:
I consider the practice a blot on our social life.
I therefore caution all wise men
That August visitors should not be admitted.

ON THE DEATH OF HIS FATHER

By Wei Wēn-ti, son of Ts'ao Ts'ao, who founded the dynasty of Wei, and died in A.D. 220. (The poem has been wrongly attributed to Han Wēn-ti, died 157 B.C.)

I LOOK up and see / his curtains and bed:
I look down and examine / his table and mat.
The things are there / just as before.
But the man they belonged to / is not there.
His spirit suddenly / has taken flight
And left me behind / far away.
To whom shall I look / on whom rely?
My tears flow / in an endless stream.
"Yu, yu" / cry the wandering deer
As they carry fodder / to their young in the wood.
Flap, flap / fly the birds
As they carry their little ones / back to the nest.
I alone / am desolate
Dreading the days / of our long parting:
My grieving heart's / settled pain
No one else / can understand.
There is a saying / among people
"Sorrow makes us / grow old."
Alas, alas / for my white hairs!
All too early / they have come!
Long wailing, / long sighing
My thoughts are fixed on my sage parent.
They say the good / live long:
Then why was *he* / not spared?

THE CAMPAIGN AGAINST WU

TWO POEMS

By Wei Wēn-ti (A.D. 188-227)

(1)

My charioteer hastens to yoke my carriage,
For I must go on a journey far away.
"Where are you going on your journey far away?"
To the land of Wu where my enemies are.
But I must ride many thousand miles,
Beyond the Eastern Road that leads to Wu.
Between the rivers bitter winds blow,
Swiftly flow the waters of Huai and Ssŭ.
I want to take a skiff and cross these rivers,
But alas for me, where shall I find a boat?
To sit idle is not my desire:
Gladly enough would I go to my country's aid.

(2)

(*He abandons the campaign*)

In the North-west there is a floating cloud
Stretched on high, like a chariot's canvas-awning.
Alas that I was born in these times,
To be blown along like a cloud puffed by the wind!
It has blown me away far to the South-east,
On and on till I came to Wu-hui.
Wu-hui is not my country:
Why should I go on staying and staying here?
I will give it up and never speak of it again,—
This being abroad and always living in dread.

THE RUINS OF LO-YANG

By Ts'ao Chih (A.D. 192-233), third son of Ts'ao Ts'ao. He was a great favourite with his father till he made a mistake in a campaign. In this poem he returns to look at the ruins of Lo-yang, where he used to live. It had been sacked by Tung Cho.

I CLIMB to the ridge of Pei Mang Mountain
And look down on the city of Lo-yang.
In Lo-yang how still it is!
Palaces and houses all burnt to ashes.
Walls and fences all broken and gaping,
Thorns and brambles shooting up to the sky.
I do not see the old old-men:
I only see the new young men.
I turn aside, for the straight road is lost:
The fields are overgrown and will never be ploughed again.
I have been away such a long time
That I do not know which street is which.
How sad and ugly the empty moors are!
A thousand miles without the smoke of a chimney.
I think of the house I lived in all those years:
I am heart-tied and cannot speak.

The above poem vaguely recalls a famous Anglo-Saxon fragment which I will make intelligible by semi-translation:

"Wondrous was the wall-stone,
Weirdly[1] broken;
Burgh-steads bursten,
Giants' work tumbleth,
Roofs are wrenched,
Towers totter,

[1] By Fate.

Bereft of rune-gates.
Smoke is on the plaster,
Scarred the shower-burghs,
Shorn and shattered,
By eld under-eaten.
Earth's grip haveth
Wealders[1] and workmen."

THE COCK-FIGHT

By Ts'ao Chih

Our wandering eyes are sated with the dancer's skill,
Our ears are weary with the sound of "kung" and "shang."[2]
Our host is silent and sits doing nothing:
All the guests go on to places of amusement.

. . . . . .

On long benches the sportsmen sit ranged
Round a cleared room, watching the fighting-cocks.
The gallant birds are all in battle-trim:
They raise their tails and flap defiantly.
Their beating wings stir the calm air:
Their angry eyes gleam with a red light.
Where their beaks have struck, the fine feathers are scattered:
With their strong talons they wound again and again.
Their long cries enter the blue clouds;
Their flapping wings tirelessly beat and throb.
"Pray God the lamp-oil lasts a little longer,
Then I shall not leave without winning the match!"

[1] Rulers. [2] Notes of the scale.

A VISION

By Ts'ao Chih

In the Nine Provinces there is not room enough:
I want to soar high among the clouds,
And, far beyond the Eight Limits of the compass,
Cast my gaze across the unmeasured void.
I will wear as my gown the red mists of sunrise,
And as my skirt the white fringes of the clouds:
My canopy—the dim lustre of Space:
My chariot—six dragons mounting heavenward:
And before the light of Time has shifted a pace
Suddenly stand upon the World's blue rim.
  The doors of Heaven swing open,
The double gates shine with a red light.
I roam and linger in the palace of Wēn-ch'ang,[1]
I climb up to the hall of T'ai-wei.[1]
The Lord God lies at his western lattice:
And the lesser Spirits are together in the eastern gallery.
They wash me in a bath of rainbow-spray
And gird me with a belt of jasper and rubies.
I wander at my ease gathering divine herbs:
I bend down and touch the scented flowers.
Wang-tzŭ[2] gives me drugs of long-life
And Hsien-mēn hands me strange potions.
By the partaking of food I evade the rites of Death:
My span is extended to the enjoyment of life everlasting.

[1] Stars. [2] Immortals.

THE CURTAIN OF THE WEDDING BED

By Liu Hsün's wife (third century A.D.).

After she had been married to him for a long while, General Liu Hsün sent his wife back to her home, because he had fallen in love with a girl of the Ssu-ma family.

FLAP, flap, you curtain in front of our bed!
I hung you there to screen us from the light of day.
I brought you with me when I left my father's house;
Now I am taking you back with me again.
I will fold you up and lay you flat in your box.
Curtain—shall I ever take you out again?

REGRET

By Yüan Chi (A.D. 210-263)

WHEN I was young I learnt fencing
And was better at it than Crooked Castle.[1]
My spirit was high as the rolling clouds
And my fame resounded beyond the World.
I took my sword to the desert sands,
I drank my horse at the Nine Moors.
My flags and banners flapped in the wind,
And nothing was heard but the song of my drums.

War and its travels have made me sad,
And a fierce anger burns within me:
It 's thinking of how I've wasted my time
That makes this fury tear my heart.

[1] A famous general.

TAOIST SONG

By Chi K'ang (A.D. 223-262)

I WILL cast out Wisdom and reject Learning.
My thoughts shall wander in the Great Void (*bis*).
Always repenting of wrongs done
Will never bring my heart to rest.
I cast my hook in a single stream;
But my joy is as though I possessed a Kingdom.
I loose my hair and go singing;
To the four frontiers men join in my refrain.
This is the purport of my song:
"My thoughts shall wander in the Great Void."

A GENTLE WIND

By Fu Hsüan (died A.D. 278)

A GENTLE wind fans the calm night:
A bright moon shines on the high tower.
A voice whispers, but no one answers when I call:
A shadow stirs, but no one comes when I beckon,
The kitchen-man brings in a dish of lentils:
Wine is there, but I do not fill my cup.
Contentment with poverty is Fortune's best gift:
Riches and Honour are the handmaids of Disaster.
Though gold and gems by the world are sought and prized,
To me they seem no more than weeds or chaff.

WOMAN

By Fu Hsüan

How sad it is to be a woman!
Nothing on earth is held so cheap.
Boys standing leaning at the door
Like Gods fallen out of Heaven.
Their hearts brave the Four Oceans,
The wind and dust of a thousand miles.
No one is glad when a girl is born:
By *her* the family sets no store.
When she grows up, she hides in her room
Afraid to look a man in the face.
No one cries when she leaves her home—
Sudden as clouds when the rain stops.
She bows her head and composes her face,
Her teeth are pressed on her red lips:
She bows and kneels countless times.
She must humble herself even to the servants.
His love is distant as the stars in Heaven,
Yet the sunflower bends toward the sun.
Their hearts more sundered than water and fire—
A hundred evils are heaped upon her.
Her face will follow the years' changes:
Her lord will find new pleasures.
They that were once like substance and shadow
Are now as far as Hu from Ch'in.[1]
Yet Hu and Ch'in shall sooner meet
Than they whose parting is like Ts'an and Ch'ēn.[2]

[1] Two lands. [2] Two stars.

DAY DREAMS

By Tso Ssŭ (third century A.D.)

WHEN I was young I played with a soft brush
And was passionately devoted to reading all sorts of books.
In prose I made Chia I my standard:
In verse I imitated Ssŭ-ma Hsiang-ju.
But then the arrows began singing at the frontier.
And a winged summons came flying to the City.
Although arms were not my profession,
I had once read Jang-Chü's war-book.
I shouted aloud and my cries rent the air:
I felt as though Tung Wu were already annihilated.
The scholar's knife cuts best at its first use
And my dreams hurried on to the completion of my plan.
I wanted at a stroke to clear the Yang-tze and Hsiang,
And at a glance to quell the Tibetans and Hu.
When my task was done, I should not accept a barony,
But refusing with a bow, retire to a cottage in the country.

THE SCHOLAR IN THE NARROW STREET

By Tso Ssŭ

FLAP, flap, the captive bird in the cage
Beating its wings against the four corners.
Depressed, depressed the scholar in the narrow street:
Clasping a shadow, he dwells in an empty house.

When he goes out, there is nowhere for him to go:
Bunches and brambles block up his path.
He composes a memorial, but it is rejected and unread,
He is left stranded, like a fish in a dry pond.
Without—he has not a single farthing of salary:
Within—there is not a peck of grain in his larder.
His relations upbraid him for his lack of success:
His friends and callers daily decrease in number.
Su Ch'in used to go preaching in the North
And Li Ssŭ sent a memorandum to the West.
I once hoped to pluck the fruits of life:
But now alas, they are all withered and dry.
Though one drinks at a river, one cannot drink more than
a bellyful;
Enough is good, but there is no use in satiety.
The bird in a forest can perch but on one bough,
And this should be the wise man's pattern.

THE DESECRATION OF THE HAN TOMBS

By Chang Tsai (third century A.D.)

At Pei-mang how they rise to Heaven,
Those high mounds, four or five in the fields!
What men lie buried under these tombs?
All of them were Lords of the Han world.
"Kung" and "Wēn"[1] gaze across at each other:
The Yüan mound is all grown over with weeds.
When the dynasty was falling, tumult and disorder arose,

[1] Names of two tombs.

Thieves and robbers roamed like wild beasts.
Of earth[1] they have carried away more than one handful,
They have gone into vaults and opened the secret doors.
Jewelled scabbards lie twisted and defaced:
The stones that were set in them, thieves have carried away,
The ancestral temples are hummocks in the ground:
The walls that went round them are all levelled flat.
Over everything the tangled thorns are growing:
A herd-boy pushes through them up the path.
Down in the thorns rabbits have made their burrows:
The weeds and thistles will never be cleared away.
Over the tombs the ploughshare will be driven
And peasants will have their fields and orchards there.
They that were once lords of a thousand hosts
Are now become the dust of the hills and ridges.
I think of what Yün-mēn[2] said
And am sorely grieved at the thought of "then" and "now."

BEARER'S SONG

By Miu Hsi (died A.D. 245). *Cf.* the "Han Burial Songs," p. 38.

When I was alive, I wandered in the streets of the Capital:
Now that I am dead, I am left to lie in the fields.
In the morning I drove out from the High Hall:

[1] In the early days of the dynasty a man stole a handful of earth from the imperial tombs, and was executed by the police. The emperor was furious at the lightness of the punishment.

[2] Yün-mēn said to Mēng Ch'ang-chün (died 279 B.C.), "Does it not grieve you to think that after a hundred years this terrace will be cast down and this pond cleared away?" Mēng Ch'ang-chün wept.

In the evening I lodged beneath the Yellow Springs.[1]
When the white sun had sunk in the Western Chasm
I hung up my chariot and rested my four horses.
Now, even the mighty Maker of All
Could not bring the life back to my limbs.
Shape and substance day by day will vanish:
Hair and teeth will gradually fall away.
Forever from of old men have been so:
And none born can escape this thing.

THE VALLEY WIND

By Lu Yün (fourth century A.D.)

LIVING in retirement beyond the World,
Silently enjoying isolation,
I pull the rope of my door tighter
And stuff my window with roots and ferns.
My spirit is tuned to the Spring-season:
At the fall of the year there is autumn in my heart.
Thus imitating cosmic changes
My cottage becomes a Universe.

[1] Hades.

CHAPTER III

POEMS BY T'AO CH'IEN

(1)

SHADY, shady the wood in front of the Hall:
At midsummer full of calm shadows.
The south wind follows summer's train:
With its eddying puffs it blows open my coat.
I am free from ties and can live a life of retirement.
When I rise from sleep, I play with books and harp.
The lettuce in the garden still grows moist:
Of last year's grain there is always plenty left.
Self-support should maintain strict limits:
More than enough is not what I want.
I grind millet and make good wine:
When the wine is heated, I pour it out for myself.
My little children are playing at my side,
Learning to talk, they babble unformed sounds.
These things have made me happy again
And I forget my lost cap of office.
Distant, distant I gaze at the white clouds:
With a deep yearning I think of the Sages of Antiquity.

(2)

IN the quiet of the morning I heard a knock at my door:
I threw on my clothes and opened it myself.

I asked who it was who had come so early to see me:
He said he was a peasant, coming with good intent.
He brought a present of wine and rice-soup,
Believing that I had fallen on evil days.
"You live in rags under a thatched roof
And seem to have no desire for a better lot.
The rest of mankind have all the same ambitions:
You, too, must learn to wallow in their mire."
"Old man, I am impressed by what you say,
But my soul is not fashioned like other men's.
To drive in their rut I might perhaps learn:
To be untrue to myself could only lead to muddle.
Let us drink and enjoy together the wine you have brought:
For my course is set and cannot now be altered."

(3)

A LONG time ago
I went on a journey,
Right to the corner
Of the Eastern Ocean.
The road there
Was long and winding,
And stormy waves
Barred my path.
What made me
Go this way?
Hunger drove me
Into the World.
I tried hard
To fill my belly:
And even a little

Seemed a lot.
But this was clearly
A bad bargain,
So I went home
And lived in idleness.

(4)
SUBSTANCE, SHADOW, AND SPIRIT

HIGH and low, wise and simple, all busily hoard up the moments of life. How greatly they err!

Therefore I have to the uttermost exposed the bitterness both of Substance and Shadow, and have made Spirit show how, by following Nature, we may dissolve this bitterness.

Substance speaks to Shadow:

Heaven and Earth exist for ever:
Mountains and rivers never change.
But herbs and trees in perpetual rotation
Are renovated and withered by the dews and frosts:
And Man the wise, Man the divine—
Shall he alone escape this law?
Fortuitously appearing for a moment in the World
He suddenly departs, never to return.
How can he know that the friends he has left
Are missing him and thinking of him?
Only the things that he used remain;
They look upon them and their tears flow.
Me no magical arts can save,
Though you may hope for a wizard's aid.

I beg you listen to this advice—
When you can get wine, be sure to drink it.

Shadow replies:

There is no way to preserve life.
Drugs of Immortality are instruments of folly.
I would gladly wander in Paradise,
But it is far away and there is no road.
Since the day that I was joined to you
We have shared all our joys and pains.
While you rested in the shade, I left you a while:
But till the end we shall be together.
Our joint existence is impermanent:
Sadly together we shall slip away.
That when the body decays Fame should also go
Is a thought unendurable, burning the heart.
Let us strive and labour while yet we may
To do some deed that men will praise.
Wine may in truth dispel our sorrow,
But how compare it with lasting Fame?

Spirit expounds:

God can only set in motion:
He cannot control the things he has made.
Man, the second of the Three Orders,
Owes his precedence to Me.
Though I am different from you,
We were born involved in one another:
Nor by any means can we escape
The intimate sharing of good and ill.
The Three Emperors were saintly men,
Yet to-day—where are they?

P'ēng[1] lived to a great age,
Yet he went at last, when he longed to stay.
And late or soon, all go:
Wise and simple have no reprieve.
Wine may bring forgetfulness,
But does it not hasten old-age?
If you set your heart on noble deeds,
How do you know that any will praise you?
By all this thinking you do Me injury:
You had better go where Fate leads—
Drift on the Stream of Infinite Flux,
Without joy, without fear:
When you must go—then go,
And make as little fuss as you can.

(5)

CHILL and harsh the year draws to its close:
In my cotton dress I seek sunlight on the porch.
In the southern orchard all the leaves are gone:
In the north garden rotting boughs lie heaped.
I empty my cup and drink it down to the dregs:
I look towards the kitchen, but no smoke rises.
Poems and books lie piled beside my chair:
But the light is going and I shall not have time to read
them.
My life here is not like the Agony in Ch'ēn,[2]
But often I have to bear bitter reproaches.
Let me then remember, to calm my heart's distress,
That the Sages of old were often in like case.

1 The Chinese Methuselah.
2 Confucius was maltreated in Ch'ēn.

(6)

BLAMING SONS

(AN APOLOGY FOR HIS OWN DRUNKENNESS)

WHITE hair covers my temples,
I am wrinkled and seared beyond repair,
And though I have got five sons,
They all hate paper and brush.
A-shu is eighteen:
For laziness there is none like him.
A-hsüan does his best,
But really loathes the Fine Arts.
Yung-tuan is thirteen,
But does not know "six" from "seven."[1]
T'ung-tzŭ in his ninth year
Is only concerned with things to eat.
If Heaven treats me like this,
What can I do but fill my cup?

(7)

I BUILT my hut in a zone of human habitation,
Yet near me there sounds no noise of horse or coach.
Would you know how that is possible?
A heart that is distant creates a wilderness round it.
I pluck chrysanthemums under the eastern hedge,
Then gaze long at the distant summer hills.
The mountain air is fresh at the dusk of day:
The flying birds two by two return.
In these things there lies a deep meaning;
Yet when we would express it, words suddenly fail us.

[1] Written in Chinese with two characters very easy to distinguish.

(8)

MOVING HOUSE

My old desire to live in the Southern Village
Was not because I had taken a fancy to the house.
But I heard it was a place of simple-minded men
With whom it were a joy to spend the mornings and evenings.
Many years I had longed to settle here:
Now at last I have managed to move house.
I do not mind if my cottage is rather small
So long as there 's room enough for bed and mat.
Often and often the neighbours come to see me
And with brave words discuss the things of old.
Rare writings we read together and praise:
Doubtful meanings we examine together and settle.

(9)

RETURNING TO THE FIELDS

When I was young, I was out of tune with the herd:
My only love was for the hills and mountains.
Unwitting I fell into the Web of the World's dust
And was not free until my thirtieth year.
The migrant bird longs for the old wood:
The fish in the tank thinks of its native pool.
I had rescued from wildness a patch of the Southern Moor
And, still rustic, I returned to field and garden.
My ground covers no more than ten acres:
My thatched cottage has eight or nine rooms.
Elms and willows cluster by the eaves:
Peach trees and plum trees grow before the Hall.

Hazy, hazy the distant hamlets of men.
Steady the smoke of the half-deserted village,
A dog barks somewhere in the deep lanes,
A cock crows at the top of the mulberry tree.
At gate and courtyard—no murmur of the World's dust:
In the empty rooms—leisure and deep stillness.
Long I lived checked by the bars of a cage:
Now I have turned again to Nature and Freedom.

(10)

READING THE BOOK OF HILLS AND SEAS

In the month of June the grass grows high
And round my cottage thick-leaved branches sway.
There is not a bird but delights in the place where it rests:
And I too—love my thatched cottage.
I have done my ploughing:
I have sown my seed.
Again I have time to sit and read my books.
In the narrow lane there are no deep ruts:
Often my friends' carriages turn back.
In high spirits I pour out my spring wine
And pluck the lettuce growing in my garden.
A gentle rain comes stealing up from the east
And a sweet wind bears it company.
My thoughts float idly over the story of King Chou
My eyes wander over the pictures of Hills and Seas.
At a single glance I survey the whole Universe.
He will never be happy, whom such pleasures fail to please!

(11)
FLOOD

THE lingering clouds, rolling, rolling,
And the settled rain, dripping, dripping,
In the Eight Directions—the same dusk.
The level lands—one great river.
Wine I have, wine I have:
Idly I drink at the eastern window.
Longingly—I think of my friends,
But neither boat nor carriage comes.

(12)
NEW CORN

SWIFTLY the years, beyond recall.
Solemn the stillness of this fair morning.
I will clothe myself in spring-clothing
And visit the slopes of the Eastern Hill.
By the mountain-stream a mist hovers,
Hovers a moment, then scatters.
There comes a wind blowing from the south
That brushes the fields of new corn.

CHAPTER IV

INVITING GUESTS

By Ch'ēng-kung Sui (died A.D. 273)

I SENT out invitations
To summon guests.
I collected together
All my friends.
Loud talk
And simple feasting:
Discussion of philosophy,
Investigation of subtleties.
Tongues loosened
And minds at one.
Hearts refreshed
By discharge of emotion!

CLIMBING A MOUNTAIN

By Tao-yün (*circa* A.D. 400), wife of General Wang Ning-chih. The general was so stupid that she finally deserted him.

HIGH rises the Eastern Peak
Soaring up to the blue sky.

Among the rocks—an empty hollow,
Secret, still, mysterious!
Uncarved and unhewn,
Screened by nature with a roof of clouds.
Times and Seasons, what things are you
Bringing to my life ceaseless change?
I will lodge for ever in this hollow
Where Springs and Autumns unheeded pass.

SAILING HOMEWARD

By Chan Fang-shēng (fourth century A.D.)

Cliffs that rise a thousand feet
Without a break,
Lake that stretches a hundred miles
Without a wave,
Sands that are white through all the year,
Without a stain,
Pine-tree woods, winter and summer
Ever-green,
Streams that for ever flow and flow
Without a pause,
Trees that for twenty thousand years
Your vows have kept,
You have suddenly healed the pain of a traveller's heart,
And moved his brush to write a new song.

FIVE "TZŬ-YEH" SONGS

At the time when blossoms
Fall from the cherry-tree:
On a day when yellow birds
Hovered in the branches—
You said you must stop,
Because your horse was tired:
I said I must go,
Because my silkworms were hungry.

All night I could not sleep
Because of the moonlight on my bed.
I kept on hearing a voice calling:
Out of Nowhere, Nothing answered "yes."

I will carry my coat and not put on my belt;
With unpainted eyebrows I will stand at the front window.
My tiresome petticoat keeps on flapping about;
If it opens a little, I shall blame the spring wind.

I heard my love was going to Yang-chou
And went with him as far as Ch'u-shan.
For a moment when you held me fast in your outstretched arms
I thought the river stood still and did not flow.

I have brought my pillow and am lying at the northern window,
So come to me and play with me awhile.
With so much quarrelling and so few kisses
How long do you think our love can last?

THE LITTLE LADY OF CH'ING-HSI

(A CHILDREN'S SONG)

HER door opened on the white water
Close by the side of the timber bridge:
That 's where the little lady lived
All alone without a lover.

PLUCKING THE RUSHES

(A BOY AND GIRL ARE SENT TO GATHER RUSHES FOR THATCHING)

Anon. (fourth century)

GREEN rushes with red shoots,
Long leaves bending to the wind—
You and I in the same boat
Plucking rushes at the Five Lakes.
We started at dawn from the orchid-island:
We rested under the elms till noon.
You and I plucking rushes
Had not plucked a handful when night came!

BALLAD OF THE WESTERN ISLAND IN THE NORTH COUNTRY

" SEEING the plum-tree I thought of the Western Island
And I plucked a branch to send to the North Country.

I put on my dress of apricot-yellow silk
And bound up my hair black as the crow's wing.
But which is the road that leads to the Western Island?
I'll ask the man at the ferry by the Bridge of Boats.
But the sun is sinking and the orioles flying home:
And the wind is blowing and sighing in the walnut-tree.
I'll stand under the tree just beside the gate:
I'll stand by the door and show off my enamelled hair-pins."
She's opened the gate, but her lover has not come:
She's gone out at the gate to pluck red lotus.
As she plucks the lotus on the southern dyke in autumn,
The lotus flowers stand higher than a man's head.
She bends down—and plays with the lotus seeds,
The lotus seeds are green like the lake-water.
She gathers the flowers and puts them into her gown—
The lotus-bud that is red all through.
She thinks of her lover, her lover that does not come:
She looks up and sees the wild geese flying—
The Western Island is full of wild geese.
To look for her lover she climbs the Blue Tower.
The tower is high: she looks, but cannot see:
All day she leans on the balcony rails.
The rail is twisted into a twelve-fold pattern.
She lets fall her hand white like the colour of jade.
She rolls up the awning, she sees the wide sky,
And the sea-water waving its vacant blue.
"The sea shall carry my dreams far away,
So that you shall be sorry at last for my sorrow.
If the South wind—only knew my thoughts
It would blow my dreams till they got to the Western
    Island."

SONG

By Tsang Chih (sixth century)

I WAS brought up under the Stone Castle:
My window opened on to the castle tower.
In the castle were beautiful young men
Who waved to me as they went in and out.

SONG OF THE MEN OF CHIN-LING

(MARCHING BACK INTO THE CAPITAL)

By Hsieh T'iao (fifth century A.D.)

CHIANG-NAN is a glorious and beautiful land,
And Chin-ling an exalted and kingly province!
The green canals of the city stretch on and on
And its high towers stretch up and up.
Flying gables lean over the bridle-road:
Drooping willows cover the Royal Aqueduct.
Shrill flutes sing by the coach's awning,
And reiterated drums bang near its painted wheels.
The names of the deserving shall be carved on the Cloud
Terrace.[1]
And for those who have done valiantly rich reward awaits.

[1] The Record Office.

THE SCHOLAR RECRUIT

By Pao Chao (died A.D. 466)

Now late
I follow Time's Necessity:[1]
Mounting a barricade I pacify remote tribes.
Discarding my sash I don a coat of rhinoceros-skin:
Rolling up my skirts I shoulder a black bow.
Even at the very start my strength fails:
What will become of me before it 's all over?

THE RED HILLS

By Pao Chao

RED hills lie athwart us as a menace in the west,
And fiery mountains glare terrible in the south.
The body burns, the head aches and throbs:
If a bird light here, its soul forthwith departs.
Warm springs
Pour from cloudy pools
And hot smoke issues between the rocks.
The sun and moon are perpetually obscured:
The rain and dew never stay dry.
There are red serpents a hundred feet long,
And black snakes ten girths round.
The sand-spitters shoot their poison at the sunbeams:
The flying insects are ill with the shifting glare.

[1] *I.e.*, "enlist."

The hungry monkeys dare not come down to eat:
The morning birds dare not set out to fly.
At the Ching river many die of poison:
Crossing the Lu one is lucky if one is only ill.
Our living feet walk on dead ground:
Our high wills surmount the snares of Fate.
The Spear-boat General[1] got but little honour:
The Wave-subduer[2] met with scant reward.
If our Prince still grudges the things that are easy to give,[3]
Can he hope that his soldiers will give what is hardest to give?[4]

DREAMING OF A DEAD LADY

"I HEARD at night your long sighs
And knew that you were thinking of me."
As she spoke, the doors of Heaven opened
And our souls conversed and I saw her face.
She set me a pillar to rest on
And she brought me meat and drink.

I stood beside her where she lay,
But suddenly woke and she was not there:
And none knew how my soul was torn,
How the tears fell surging over my breast.

[1] Hou Yen (first century B.C.).
[2] Ma Yüan (first century A.D.).
[3] Rewards and titles.
[4] Life.

THE LIBERATOR

A POLITICAL ALLEGORY

By Wu-ti, emperor of the Liang dynasty (A.D. 464-549)

In the high trees—many doleful winds:
The ocean waters—lashed into waves.
If the sharp sword be not in your hand,
How can you hope your friends will remain many?
Do you not see that sparrow on the fence?
Seeing the hawk it casts itself into the snare.
The fowler to catch the sparrow is delighted:
The Young Man to see the sparrow is grieved.
He takes his sword and cuts through the netting:
The yellow sparrow flies away, away.
Away, away, up to the blue sky
And down again to thank the Young Man.

LO-YANG

By the Emperor Ch'ien Wēn-ti (sixth century)

A beautiful place is the town of Lo-yang:
The big streets are full of spring light.
The lads go driving out with harps in their hands:
The mulberry girls go out to the fields with their baskets.
Golden whips glint at the horses' flanks,
Gauze sleeves brush the green boughs.
Racing dawn, the carriages come home,—
And the girls with their high baskets full of fruit.

WINTER NIGHT

My bed is so empty that I keep on waking up:
As the cold increases, the night-wind begins to blow.
It rustles the curtains, making a noise like the sea:
Oh that those were waves which could carry me back to you!

THE REJECTED WIFE

By Yüan-ti (508-554). See page 15.

Entering the Hall, she meets the new wife:
Leaving the gate, she runs into her former husband.
Words stick: she does not manage to say anything:
She presses her hands together and hesitates.
Agitates moon-like fan—sheds pearl-like tears—
Realizes she loves him just as much as ever:
That her present pain will never come to an end.

PEOPLE HIDE THEIR LOVE

By Wu-ti

Who says
That it 's by my desire,
This separation, this living so far from you?
My dress still smells of the lavender you gave:
My hand still holds the letter that you sent.
Round my waist I wear a double sash:
I dream that it binds us both with a same-heart knot.
Did not you know that people hide their love,
Like a flower that seems too precious to be picked?

THE FERRY

By the Emperor Ch'ien Wēn-ti, of the Liang dynasty, who reigned during the year A.D. 500.

Of marsh-mallows my boat is made,
The ropes are lily-roots.
The pole-star is athwart the sky:
The moon sinks low.
It 's at the ferry I'm plucking lilies.
But it might be the Yellow River—
So afraid you seem of the wind and waves,
So long you tarry at the crossing.[1]

[1] A lady is waiting for her lover at the ferry which crosses a small stream. When he does not come, she bitterly suggests that he is as afraid of the little stream as though it were the Yellow River, the largest river in China.

THE WATERS OF LUNG-T'OU

(THE NORTH-WEST FRONTIER)

By Hsü Ling (A.D. 507-583)

THE road that I came by mounts eight thousand feet:
The river that I crossed hangs a hundred fathoms.
The brambles so thick that in summer one cannot pass!
The snow so high that in winter one cannot climb!
With branches that interlace Lung Valley is dark:
Against cliffs that tower one's voice beats and echoes.
I turn my head, and it seems only a dream
That I ever lived in the streets of Hsien-yang.

FLOWERS AND MOONLIGHT ON THE SPRING RIVER

By Yang-ti (605-617), emperor of the Sui dynasty

THE evening river is level and motionless—
The spring colours just open to their full.
Suddenly a wave carries the moon[1] away
And the tidal water comes with its freight of stars.[1]

[1] *I.e.*, the reflection in the water.

TCHIREK SONG

Altun (486-566 A.D.) was a Tartar employed by the Chinese in drilling their troops "after the manner of the Huns." He could not read or write. The "Yo Fu Kuang T'i" says: Kao Huan attacked Pi, king of Chou, but lost nearly half his men. Kao Huan fell ill of sadness and Pi, to taunt him, sent out a proclamation, which said:

Kao Huan, that son of a mouse
Dared to attack King Pi.
But at the first stroke of sword and bow,
The aggressor's plot recoiled on himself.

When this reached Kao Huan's ears, he sat up in bed and tried to comfort his officers. All the nobles were summoned to his room, and Altun was asked to sing them a song about Tchirek, his native land. He sang:

TCHIREK River
Lies under the Dark Mountains:
Where the sky is like the sides of a tent
Stretched down over the Great Steppe.
The sky is gray, gray:
And the steppe wide, wide:
Over grass that the wind has battered low
Sheep and oxen roam.

"Altun" means "gold" in Tartar. No one could teach him to write the Chinese character for *gold*, till at last some one said: "Draw the roof of your house and then put a few strokes underneath." He thus learnt, in a rough fashion, to write his own name.

CHAPTER V

BUSINESS MEN

By Ch'ĕn Tzŭ-ang (A.D. 656-698)

BUSINESS men boast of their skill and cunning
But in philosophy they are like little children.
Bragging to each other of successful depredations
They neglect to consider the ultimate fate of the body.
What should they know of the Master of Dark Truth
Who saw the wide world in a jade cup,
By illumined conception got clear of Heaven and Earth:
On the chariot of Mutation entered the Gate of Immutability?

TELL ME NOW

By Wang Chi (*circa* A.D. 700)

"TELL me now, what should a man want
But to sit alone, sipping his cup of wine?"
I should like to have visitors come and discuss philosophy
And not to have the tax-collector coming to collect taxes:
My three sons married into good families
And my five daughters wedded to steady husbands.
Then I could jog through a happy five-score years
And, at the end, need no Paradise.

ON GOING TO A TAVERN

By Wang Chi

THESE days, continually fuddled with drink,
I fail to satisfy the appetites of the soul.
But seeing men all behaving like drunkards,[1]
How can I alone remain sober?

STONE FISH LAKE

By Yüan Chieh (flourished *circa* A.D. 740-770).

Yüan Chieh, a contemporary of Li Po, has not hitherto been mentioned in any European book. "His subjects were always original, but his poems are seldom worth quoting," is a Chinese opinion of him.

I LOVED you dearly, Stone Fish Lake,
With your rock-island shaped like a swimming fish!
On the fish's back is the Wine-cup Hollow
And round the fish,—the flowing waters of the Lake.
The boys on the shore sent little wooden ships,
Each made to carry a single cup of wine.
The island-drinkers emptied the liquor-boats
And set their sails and sent them back for more.
On the shores of the Lake were jutting slabs of rock
And under the rocks there flowed an icy stream.
Heated with wine, to rinse our mouths and hands
In those cold waters was a joy beyond compare!

Of gold and jewels I have not any need;
For Caps and Coaches I do not care at all.
But I wish I could sit on the rocky banks of the Lake
For ever and ever staring at the Stone Fish.

[1] Written during the war which preceded the T'ang dynasty.

CIVILIZATION

By Yüan Chieh

To the south-east—three thousand leagues—
The Yüan and Hsiang form into a mighty lake.
Above the lake are deep mountain valleys,
And men dwelling whose hearts are without guile.
Gay like children, they swarm to the tops of the trees;
And run to the water to catch bream and trout.
Their pleasures are the same as those of beasts and birds;
They put no restraint either on body or mind.
Far I have wandered throughout the Nine Lands;
Wherever I went such manners had disappeared.
I find myself standing and wondering, perplexed,
Whether Saints and Sages have really done us good.

A PROTEST IN THE SIXTH YEAR OF CH'IEN FU (A.D. 879)

By Ts'ao Sung (flourished *circa* A.D. 870-920)

THE hills and rivers of the lowland country
  You have made your battle-ground.
How do you suppose the people who live there
  Will procure "firewood and hay"?[1]

[1] The necessaries of life.

Do not let me hear you talking together
  About titles and promotions;
For a single general's reputation
  Is made out of ten thousand corpses.

ON THE BIRTH OF HIS SON

By Su Tung-p'o (A.D. 1036-1101)

Families, when a child is born
Want it to be intelligent.
I, through intelligence,
Having wrecked my whole life,
Only hope the baby will prove
Ignorant and stupid.
Then he will crown a tranquil life
By becoming a Cabinet Minister.

THE PEDLAR OF SPELLS

By Lu Yu (A.D. 1125-1209)

An old man selling charms in a cranny of the town wall.
He writes out spells to bless the silkworms and spells to
    protect the corn.
With the money he gets each day he only buys wine.
But he does not worry when his legs get wobbly,
For he has a boy to lean on.

BOATING IN AUTUMN

By Lu Yu

Away and away I sail in my light boat;
My heart leaps with a great gust of joy.
Through the leafless branches I see the temple in the wood;
Over the dwindling stream the stone bridge towers.
Down the grassy lanes sheep and oxen pass;
In the misty village cranes and magpies cry.

Back in my home I drink a cup of wine
And need not fear the greed[1] of the evening wind.

THE HERD-BOY

By Lu Yu

In the southern village the boy who minds the ox
With his naked feet stands on the ox's back.
Through the hole in his coat the river wind blows;
Through his broken hat the mountain rain pours.
On the long dyke he seemed to be far away;
In the narrow lane suddenly we were face to face.

The boy is home and the ox is back in its stall;
And a dark smoke oozes through the thatched roof.

[1] Which "eats" men.

HOW I SAILED ON THE LAKE TILL I CAME TO THE EASTERN STREAM

By Lu Yu

Of Spring water,—thirty or forty miles:
In the evening sunlight,—three or four houses.
Youths and boys minding geese and ducks:
Women and girls tending mulberries and hemp.
The place,—remote: their coats and scarves old:
The year,—fruitful: their talk and laughter gay.
The old wanderer moors his flat boat
And staggers up the bank to pluck wistaria flowers.

A SEVENTEENTH-CENTURY CHINESE POEM

Ch'ēn Tzŭ-lung was born in 1607. He became a soldier, and in 1637 defeated the rebel, Hsü Tu. After the suicide of the last Ming emperor, he offered his services to the Ming princes who were still opposing the Manchus. In 1647 he headed a conspiracy to place the Ming prince Lu on the throne. His plans were discovered and he was arrested by Manchu troops. Escaping their vigilance for a moment, he leapt into a river and was drowned.

The following song describes the flight of a husband and wife from a town menaced by the advancing Manchus. They find the whole country-side deserted.

THE LITTLE CART

The little cart jolting and banging through the yellow haze
of dusk.
The man pushing behind: the woman pulling in front.
They have left the city and do not know where to go.

"Green, green, those elm-tree leaves: *they* will cure my hunger,
If only we could find some quiet place and sup on them together."

The wind has flattened the yellow mother-wort:
Above it in the distance they see the walls of a house.
"*There* surely must be people living who'll give you something to eat."
They tap at the door, but no one comes: they look in, but the kitchen is empty.
They stand hesitating in the lonely road and their tears fall like rain.

PART II

Po Chü-i

(A.D. 772—846)

INTRODUCTION

PO CHÜ-I was born at T'ai-yüan in Shansi. Most of his childhood was spent at Jung-yang in Honan. His father was a second-class Assistant Department Magistrate. He tells us that his family was poor and often in difficulties.

He seems to have settled permanently at Ch'ang-an in 801. This town, lying near the north-west frontier, was the political capital of the Empire. In its situation it somewhat resembled Madrid. Lo-yang, the Eastern city, owing to its milder climate and more accessible position, became, like Seville in Spain, a kind of *social* capital.

Soon afterwards he met Yüan Chēn, then aged twenty-two, who was destined to play so important a part in his life. Five years later, during a temporary absence from the city, he addressed to Yüan the following poem:

Since I left my home to seek official state
Seven years I have lived in Ch'ang-an.
What have I gained? Only you, Yüan;
So hard it is to bind friendships fast.
We have roamed on horseback under the flowering trees;
We have walked in the snow and warmed our hearts with wine.
    We have met and parted at the Western Gate
And neither of us bothered to put on Cap or Belt.
We did not go up together for Examination;
We were not serving in the same Department of State.
The bond that joined us lay deeper than outward things;
The rivers of our souls spring from the same well!

Of Yüan's appearance at this time we may guess something from a picture which still survives in copy; it shows him, a youthful and elegant figure, visiting his cousin Ts'ui Ying-ying, who was a lady-in-waiting at Court.[1] At this period of his life Po made friends with difficulty, not being, as he tells us "a master of such accomplishments as caligraphy, painting, chess or gambling, which tend to bring men together in pleasurable intercourse." Two older men, T'ang Ch'ü and Tēng Fang, liked his poetry and showed him much kindness; another, the politician K'ung T'an, won his admiration on public grounds. But all three died soon after he got to know them. Later he made three friends with whom he maintained a lifelong intimacy: the poet Liu Yü-hsi (called Mēng-tē), and the two officials Li Chien and Ts'ui Hsuan-liang. In 805 Yüan Chēn was banished for provocative behaviour towards a high official. The T'ang History relates the episode as follows: "Yüan was staying the night at the Fu-shui Inn; just as he was preparing to go to sleep in the Main Hall, the court-official Li Shih-yüan also arrived. Yüan Chēn should have offered to withdraw from the Hall. He did not do so and a scuffle ensued. Yüan, locked out of the building, took off his shoes and stole round to the back, hoping to find another way in. Liu followed with a whip and struck him across the face."

The separation was a heavy blow to Po Chü-i. In a poem called "Climbing Alone to the Lo-yu Gardens" he says:

[1] Yüan has told the story of this intrigue in an autobiographical fragment, of which I hope to publish a translation. Upon this fragment is founded the famous fourteenth-century drama, "The Western Pavilion."

I look down on the Twelve City Streets:—
Red dust flanked by green trees!
Coaches and horsemen alone fill my eyes;
I do not see whom my heart longs to see.
K'ung T'an has died at Lo-yang;
Yüan Chēn is banished to Ching-mēn.
Of all that walk on the North-South Road
There is not one that I care for more than the rest!

In 804 on the death of his father, and again in 811 on the death of his mother, he spent periods of retirement on the Wei river near Ch'ang-an. It was during the second of these periods that he wrote the long poem (260 lines) called "Visiting the Wu-chēn Temple." Soon after his return to Ch'ang-an, which took place in the winter of 814, he fell into official disfavour. In two long memorials entitled "On Stopping the War," he had criticized the handling of a campaign against an unimportant tribe of Tartars, which he considered had been unduly prolonged. In a series of poems he had satirized the rapacity of minor officials and called attention to the intolerable sufferings of the masses.

His enemies soon found an opportunity of silencing him. In 814 the Prime Minister, Wu Yüan-hēng, was assassinated in broad daylight by an agent of the revolutionary leader Wu Yüan-chi. Po, in a memorial to the Throne, pointed out the urgency of remedying the prevailing discontent. He held at this time the post of assistant secretary to the Princes' tutor. He should not have criticized the Prime Minister (for being murdered!) until the official Censors had spoken, for he held a Palace appointment which did not carry with it the right of censorship.

His opponents also raked up another charge. His mother had met her death by falling into a well while looking at flowers. Chü-i had written two poems entitled

"In Praise of Flowers" and "The New Well." It was claimed that by choosing such subjects he had infringed the laws of Filial Piety.

He was banished to Kiukiang (then called Hsün-yang) with the rank of Sub-Prefect. After three years he was given the Governorship of Chung-chou, a remote place in Ssech'uan. On the way up the Yangtze he met Yüan Chēn after three years of separation. They spent a few days together at I-ch'ang, exploring the rock-caves of the neighbourhood.

Chung-chou is noted for its "many flowers and exotic trees," which were a constant delight to its new Governor. In the winter of 819 he was recalled to the capital and became a second-class Assistant Secretary. About this time Yüan Chēn also returned to the city.

In 821 the Emperor Mou Tsung came to the throne. His arbitrary mis-government soon caused a fresh rising in the north-west. Chü-i remonstrated in a series of memorials and was again removed from the capital—this time to be Governor of the important town of Hangchow. Yüan now held a judicial post at Ningpo and the two were occasionally able to meet.

In 824 his Governorship expired and he lived (with the nominal rank of Imperial Tutor) at the village of Li-tao-li, near Lo-yang. It was here that he took into his household two girls, Fan-su and Man-tzŭ, whose singing and dancing enlivened his retreat. He also brought with him from Hangchow a famous "Indian rock," and two cranes of the celebrated "Hua-t'ing" breed. Other amenities of his life at this time were a recipe for making sweet wine, the gift of Ch'ēn Hao-hsien; a harp-melody taught him by Ts'ui Hsuan-liang; and a song called "Autumn

Thoughts," brought by the concubine of a visitor from Ssech'uan.

In 825 he became Governor of Soochow. Here at the age of fifty-three he enjoyed a kind of second youth, much more sociable than that of thirty years before; we find him endlessly picnicking and feasting. But after two years illness obliged him to retire.

He next held various posts at the capital, but again fell ill, and in 829 settled at Lo-yang as Governor of the Province of Honan. Here his first son, A-ts'ui, was born, but died in the following year.

In 831 Yüan Chēn also died.

Henceforth, though for thirteen years he continued to hold nominal posts, he lived a life of retirement. In 832 he repaired an unoccupied part of the Hsiang-shan monastery at Lung-mēn,[1] a few miles south of Lo-yang, and lived there, calling himself the Hermit of Hsiang-shan. Once he invited to dinner eight other elderly and retired officials; the occasion was recorded in a picture entitled "The Nine Old Men at Hsiang-shan." There is no evidence that his association with them was otherwise than transient, though legend (see "Memories Concernant les Chinois" and Giles, Biographical Dictionary) has invested the incident with an undue importance. He amused himself at this time by writing a description of his daily life which would be more interesting if it were not so closely modelled on a famous memoir by T'ao Ch'ien. In the winter of 839 he was attacked by paralysis and lost the use of his left leg. After many months in bed he was again able to visit his garden, carried by Ju-man, a favourite monk.

[1] Famous for its rock-sculptures, carved in the sixth and seventh centuries.

In 842 Liu Yü-hsi, the last survivor of the four friends, and a constant visitor at the monastery, "went to wander with Yüan Chēn in Hades." The monk Ju-man also died.

The remaining years of Po's life were spent in collecting and arranging his Complete Works. Copies were presented to the principal monasteries (the "Public Libraries" of the period) in the towns with which he had been connected. He died in 846, leaving instructions that his funeral should be without pomp and that he should be buried not in the family tomb at Hsia-kuei, but by Ju-man's side in the Hsiang-shan Monastery. He desired that a posthumous title should not be awarded.

The most striking characteristic of Po Chü-i's poetry is its verbal simplicity. There is a story that he was in the habit of reading his poems to an old peasant woman and altering any expression which she could not understand. The poems of his contemporaries were mere elegant diversions which enabled the scholar to display his erudition, or the literary juggler his dexterity. Po expounded his theory of poetry in a letter to Yüan Chēn. Like Confucius, he regarded art solely as a method of conveying instruction. He is not the only great artist who has advanced this untenable theory. He accordingly valued his didactic poems far above his other work; but it is obvious that much of his best poetry conveys no moral whatever. He admits, indeed, that among his "miscellaneous stanzas" many were inspired by some momentary sensation or passing event. "A single laugh or a single sigh were rapidly translated into verse."

The didactic poems or "satires" belong to the period before his first banishment. "When the tyrants and favourites heard my Songs of Ch'in, they looked at one another and changed countenance," he boasts. Satire, in the European sense, implies *wit*; but Po's satires are as lacking in true wit as they are unquestionably full of true poetry. We must regard them simply as moral tales in verse.

In the conventional lyric poetry of his predecessors he finds little to admire. Among the earlier poems of the T'ang dynasty he selects for praise the series by Ch'ēn Tzŭ-ang, which includes "Business Men." In Li Po and Tu Fu he finds a deficiency of "fēng" and "ya." The two terms are borrowed from the Preface to the Odes. "Fēng" means "criticism of one's rulers"; "ya," "moral guidance to the masses."

"The skill," he says in the same letter, "which Tu Fu shows in threading on to his *lü-shih* a ramification of allusions ancient and modern could not be surpassed; in this he is even superior to Li Po. But, if we take the 'Press-gang' and verses like that stanza:

> At the palace doors the smell of meat and wine;
> On the road the bones of one who was frozen to death.

what a small part of his whole work it represents!"

Content, in short, he valued far above form: and it was part of his theory, though certainly not of his practice, that this content ought to be definitely moral. He aimed at raising poetry from the triviality into which it had sunk and restoring it to its proper intellectual level. It is an irony that he should be chiefly known to posterity, in China, Japan, and the West, as the author of the "Everlasting Wrong."[1] He set little store by the poem himself,

[1] Giles, "Chinese Literature," p. 169.

and, though a certain political moral might be read into it, its appeal is clearly romantic.

His other poem of sentiment, the "Lute Girl,"[1] accords even less with his stated principles. With these he ranks his *Lü-shih*; and it should here be noted that all the satires and long poems are in the old style of versification, while his lighter poems are in the strict, modern form. With his satires he classes his "reflective" poems, such as "Singing in the Mountains," "On being removed from Hsün-yang," "Pruning Trees," etc. These are all in the old style.

No poet in the world can ever have enjoyed greater contemporary popularity than Po. His poems were "on the mouths of kings, princes, concubines, ladies, plough-boys, and grooms." They were inscribed "on the walls of village-schools, temples, and ships-cabins." "A certain Captain Kao Hsia-yü was courting a dancing-girl. 'You must not think I am an ordinary dancing-girl,' she said to him, 'I can recite Master Po's "Everlasting Wrong."' And she put up her price."

But this popularity was confined to the long, romantic poems and the *Lü-shih*. "The world," writes Po to Yüan Chēn, "values highest just those of my poems which I most despise. Of contemporaries you alone have understood my satires and reflective poems. A hundred, a thousand years hence perhaps some one will come who will understand them as you have done."

The popularity of his lighter poems lasted till the Ming dynasty, when a wave of pedantry swept over China. At that period his poetry was considered vulgar, because it was not erudite; and prosaic, because it was not rhetorical.

[1] Giles, "Chinese Literature," p. 165.

Although they valued form far above content, not even the Ming critics can accuse him of slovenly writing. His versification is admitted by them to be "correct."

Caring, indeed, more for matter than for manner, he used with facility and precision the technical instruments which were at his disposal. Many of the later anthologies omit his name altogether, but he has always had isolated admirers. Yüan Mei imitates him constantly, and Chao I (died 1814) writes: "Those who accuse him of being vulgar and prosaic know nothing of poetry."

Even during his lifetime his reputation had reached Japan, and great writers like Michizane were not ashamed to borrow from him. He is still held in high repute there, is the subject of a Nō Play and has even become a kind of Shintō deity. It is significant that the only copy of his works in the British Museum is a seventeenth-century Japanese edition.

It is usual to close a biographical notice with an attemp to describe the "character" of one's subject. But I hold myself absolved from such a task; for the sixty poems which follow will enable the reader to perform it for himself.

AN EARLY LEVÉE

ADDRESSED TO CH'ĒN, THE HERMIT

AT Ch'ang-an—a full foot of snow;
A levée at dawn—to bestow congratulations on the
Emperor.
Just as I was nearing the Gate of the Silver Terrace,
After I had left the suburb of Hsin-ch'ang
On the high causeway my horse's foot slipped;
In the middle of the journey my lantern suddenly went out.
Ten leagues riding, always facing to the North;
The cold wind almost blew off my ears.
I waited for the bell outside the Five Gates;
I waited for the summons within the Triple Hall.
My hair and beard were frozen and covered with icicles;
My coat and robe—chilly like water.
Suddenly I thought of Hsien-yu Valley
And secretly envied Ch'ēn Chü-shih,
In warm bed-socks dozing beneath the rugs
And not getting up till the sun has mounted the sky.

BEING ON DUTY ALL NIGHT IN THE PALACE AND DREAMING OF THE HSIEN-YU TEMPLE

At the western window I paused from writing rescripts;
The pines and bamboos were all buried in stillness.
The moon rose and a calm wind came;
Suddenly, it was like an evening in the hills.
And so, as I dozed, I dreamed of the South West
And thought I was staying at the Hsien-yu Temple.[1]
When I woke and heard the dripping of the Palace clock
I still thought it the murmur of a mountain stream.

PASSING T'IEN-MĒN STREET IN CH'ANG-AN AND SEEING A DISTANT VIEW OF CHUNG-NAN[2] MOUNTAIN

The snow has gone from Chung-nan; spring is almost come.
Lovely in the distance its blue colours, against the brown of the streets.
A thousand coaches, ten thousand horsemen pass down the Nine Roads;
Turn his head and looks at the mountains,—not one man!

[1] Where the poet used to spend his holidays.

[2] Part of the great Nan Shan range, fifteen miles south of Ch'ang-an.

THE LETTER

Preface:—After I parted with Yüan Chēn, I suddenly dreamt one night that I saw him. When I awoke, I found that a letter from him had just arrived and, enclosed in it, a poem on the *paulovnia* flower.

We talked together in the Yung-shou Temple;
We parted to the north of the Hsin-ch'ang dyke.
Going home—I shed a few tears,
Grieving about things,—not sorry for you.
Long, long the road to Lan-t'ien;
You said yourself you would not be able to write.
Reckoning up your halts for eating and sleeping—
By this time you've crossed the Shang mountains.
Last night the clouds scattered away;
A thousand leagues, the same moonlight scene.
When dawn came, I dreamt I saw your face;
It must have been that you were thinking of me.
In my dream, I thought I held your hand
And asked you to tell me what your thoughts were.
And *you* said: "I miss you bitterly,
But there 's no one here to send to you with a letter."
When I awoke, before I had time to speak,
A knocking on the door sounded "Doong, doong!"
They came and told me a messenger from Shang-chou
Had brought a letter,—a single scroll from you!
Up from my pillow I suddenly sprang out of bed,
And threw you my clothes, all topsy-turvy.
I undid the knot and saw the letter within;
A single sheet with thirteen lines of writing.

THE LETTER

At the top it told the sorrows of an exile's heart;
At the bottom it described the pains of separation.
The sorrows and pains took up so much space
There was no room left to talk about the weather!
  But you said that when you wrote
You were staying for the night to the east of Shang-chou;
Sitting alone, lighted by a solitary candle
Lodging in the mountain hostel of Yang-Ch'ēng.
  Night was late when you finished writing,
The mountain moon was slanting towards the west.
What is it lies aslant across the moon?
A single tree of purple *paulovnia* flowers—
Paulovnia flowers just on the point of falling
Are a symbol to express "thinking of an absent friend."
Lovingly—you wrote on the back side,
To send in the letter, your "Poem of the Paulovnia Flower."
The Poem of the Paulovnia Flower has eight rhymes;
Yet these eight couplets have cast a spell on my heart.
They have taken hold of this morning's thoughts
And carried them to yours, the night you wrote your letter
The whole poem I read three times;
Each verse ten times I recite.
So precious to me are the fourscore words
That each letter changes into a bar of gold!

REJOICING AT THE ARRIVAL OF CH'ĒN HSIUNG

(*Circa* A.D. 812)

WHEN the yellow bird's note was almost stopped;
And half formed the green plum's fruit;

Sitting and grieving that spring things were over,
I rose and entered the Eastern Garden's gate.
I carried my cup and was dully drinking alone:
Suddenly I heard a knocking sound at the door.
Dwelling secluded, I was glad that someone had come;
How much the more, when I saw it was Ch'ēn Hsiung!
At ease and leisure,—all day we talked;
Crowding and jostling,—the feelings of many years.
How great a thing is a single cup of wine!
For it makes us tell the story of our whole lives.

GOLDEN BELLS

When I was almost forty
I had a daughter whose name was Golden Bells.
Now it is just a year since she was born;
She is learning to sit and cannot yet talk.
Ashamed,—to find that I have not a sage's heart:
I cannot resist vulgar thoughts and feelings.
Henceforward I am tied to things outside myself:
My only reward,—the pleasure I am getting now.
If I am spared the grief of her dying young,
Then I shall have the trouble of getting her married.
My plan for retiring and going back to the hills
Must now be postponed for fifteen years!

REMEMBERING GOLDEN BELLS

RUINED and ill,—a man of two score;
  Pretty and guileless,—a girl of three.
Not a boy,—but still better than nothing:
To soothe one's feeling,—from time to time a kiss!
There came a day,—they suddenly took her from me;
Her soul's shadow wandered I know not where.
And when I remember how just at the time she died
She lisped strange sounds, beginning to learn to talk,
Then I know that the ties of flesh and blood
Only bind us to a load of grief and sorrow.
At last, by thinking of the time before she was born,
By thought and reason I drove the pain away.
Since my heart forgot her, many days have passed
And three times winter has changed to spring.
This morning, for a little, the old grief came back,
Because, in the road, I met her foster-nurse.

ILLNESS

SAD, sad—lean with long illness;
Monotonous, monotonous—days and nights pass.
The summer trees have clad themselves in shade;
The autumn "lan"[1] already houses the dew.

[1] The epidendrum.

The eggs that lay in the nest when I took to bed
Have changed into little birds and flown away.
The worm that then lay hidden in its hole
Has hatched into a cricket sitting on the tree.
The Four Seasons go on for ever and ever:
In all Nature nothing stops to rest
Even for a moment. Only the sick man's heart
Deep down still aches as of old!

THE DRAGON OF THE BLACK POOL

A Satire

Deep the waters of the Black Pool, coloured like ink;
They say a Holy Dragon lives there, whom men have never seen.
Beside the Pool they have built a shrine; the authorities have established a ritual;
A dragon by itself remains a dragon, but men can make it a god.
Prosperity and disaster, rain and drought, plagues and pestilences—
By the village people were all regarded as the Sacred Dragon's doing.
They all made offerings of sucking-pig and poured libations of wine;
The morning prayers and evening gifts depended on a "medium's" advice

When the dragon comes, ah!
The wind stirs and sighs
Paper money thrown, ah!
Silk umbrellas waved.
When the dragon goes, ah!
The wind also—still.
Incense-fire dies, ah!
The cups and vessels are cold.[1]

Meats lie stacked on the rocks of the Pool's shore;
Wine flows on the grass in front of the shrine.
I do not know, of all those offerings, how much the Dragon eats;
But the mice of the woods and the foxes of the hills are continually drunk and sated.
Why are the foxes so lucky?
What have the sucking-pigs done,
That year by year *they* should be killed, merely to glut the foxes?
That the foxes are robbing the Sacred Dragon and eating His sucking-pig,
Beneath the nine-fold depths of His pool, does He know or not?

[1] Parody of a famous Han dynasty hymn.

THE GRAIN-TRIBUTE

Written *circa* 812, showing one of the poet's periods of retirement. When the officials come to receive his grain-tribute, he remembers that he is only giving back what he had taken during his years of office. Salaries were paid partly in kind.

THERE came an officer knocking by night at my door—
In a loud voice demanding grain-tribute.
My house-servants dared not wait till the morning,
But brought candles and set them on the barn-floor.
Passed through the sieve, clean-washed as pearls,
A whole cart-load, thirty bushels of grain.
But still they cry that it is not paid in full:
With whips and curses they goad my servants and boys.
Once, in error, I entered public life;
I am inwardly ashamed that my talents were not sufficient.
In succession I occupied four official posts;
For doing nothing,—ten years' salary!
Often have I heard that saying of ancient men
That "good and ill follow in an endless chain."
And to-day it ought to set my heart at rest
To return to others the corn in my great barn.

THE PEOPLE OF TAO-CHOU

In the land of Tao-chou
Many of the people are dwarfs;
The tallest of them never grow to more than three feet.

They were sold in the market as dwarf slaves and yearly sent to Court;
Described as "an offering of natural products from the land of Tao-chou."
A strange "offering of natural products"; I never heard of one yet
That parted men from those they loved, never to meet again!
Old men—weeping for their grandsons; mothers for their children!
One day—Yang Ch'ĕng came to govern the land;
He refused to send up dwarf slaves in spite of incessant mandates.
He replied to the Emperor "Your servant finds in the Six Canonical Books
'In offering products, one must offer what is there, and not what isn't there'
On the waters and lands of Tao-chou, among all the things that live
I only find dwarfish *people*; no dwarfish *slaves*."
The Emperor's heart was deeply moved and he sealed and sent a scroll
"The yearly tribute of dwarfish slaves is henceforth annulled."
The people of Tao-chou,
Old ones and young ones, how great their joy!
Father with son and brother with brother henceforward kept together;
From that day for ever more they lived as free men.
The people of Tao-chou
Still enjoy this gift.
And even now when they speak of the Governor

Tears start to their eyes.
And lest their children and their children's children should forget the Governor's name,
When boys are born the syllable "Yang" is often used in their forename.

THE OLD HARP

Of cord and cassia-wood is the harp compounded:
Within it lie ancient melodies.
Ancient melodies—weak and savourless,
Not appealing to present men's taste.
Light and colour are faded from the jade stops:
Dust has covered the rose-red strings.
Decay and ruin came to it long ago,
But the sound that is left is still cold and clear.
I do not refuse to play it, if you want me to:
But even if I play, people will not listen.

. . . . . .

How did it come to be neglected so?
Because of the Ch'iang flute and the Ch'in flageolet.[1]

THE HARPER OF CHAO

The singers have hushed their notes of clear song:
The red sleeves of the dancers are motionless.
Hugging his lute, the old harper of Chao
Rocks and sways as he touches the five chords.

[1] Barbarous modern instruments.

The loud notes swell and scatter abroad:
"Sa, sa," like wind blowing the rain.
The soft notes dying almost to nothing:
"Ch'ieh, ch'ieh," like the voice of ghosts talking.
Now as glad as the magpie's lucky song:
Again bitter as the gibbon's ominous cry.
His ten fingers have no fixed note:
Up and down—"kung," chih, and yü.[1]
And those who sit and listen to the tune he plays
Of soul and body lose the mastery.
And those who pass that way as he plays the tune,
Suddenly stop and cannot raise their feet.

Alas, alas that the ears of common men
Should love the modern and not love the old.
Thus it is that the harp in the green window
Day by day is covered deeper with dust.

THE FLOWER MARKET

In the Royal City spring is almost over:
Tinkle, tinkle—the coaches and horsemen pass.
We tell each other "This is the peony season":
And follow with the crowd that goes to the Flower Market.
"Cheap and dear—no uniform price:
The cost of the plant depends on the number of blossoms.
For the fine flower,—a hundred pieces of damask:
For the cheap flower,—five bits of silk.

[1] Tonic, dominant and superdominant of the ancient five-note scale.

Above is spread an awning to protect them:
Around is woven a wattle-fence to screen them.
If you sprinkle water and cover the roots with mud,
When they are transplanted, they will not lose their beauty."
Each household thoughtlessly follows the custom,
Man by man, no one realizing.
There happened to be an old farm labourer
Who came by chance that way.
He bowed his head and sighed a deep sigh:
But this sigh nobody understood.
He was thinking, "A cluster of deep-red flowers
Would pay the taxes of ten poor houses."

THE PRISONER

Written in A.D. 809

TARTARS led in chains,
Tartars led in chains!
Their ears pierced, their faces bruised—they are driven into the land of Ch'in.
The Son of Heaven took pity on them and would not have them slain.
He sent them away to the south-east, to the lands of Wu and Yüeh.
A petty officer in a yellow coat took down their names and surnames:
They were led from the city of Ch'ang-an under escort of an armed guard.

Their bodies were covered with the wounds of arrows, their bones stood out from their cheeks.
They had grown so weak they could only march a single stage a day.
In the morning they must satisfy hunger and thirst with neither plate nor cup:
At night they must lie in their dirt and rags on beds that stank with filth.
Suddenly they came to the Yangtze River and remembered the waters of Chiao.[1]
With lowered hands and levelled voices they sobbed a muffled song.
Then one Tartar lifted up his voice and spoke to the other Tartars,
"*Your* sorrows are none at all compared with *my* sorrows."
Those that were with him in the same band asked to hear his tale:
As he tried to speak the words were choked by anger.
He told them "I was born and bred in the town of Liang-yüan.[2]
In the frontier wars of Ta-li[3] I fell into the Tartars' hands.
Since the days the Tartars took me alive forty years have passed:
They put me into a coat of skins tied with a belt of rope.
Only on the first of the first month might I wear my Chinese dress.
As I put on my coat and arranged my cap, how fast the tears flowed!
I made in my heart a secret vow I would find a way home:

[1] In Turkestan. [2] North of Ch'ang-an.
[3] The period Ta-li, A.D. 766-780.

I hid my plan from my Tartar wife and the children she
had borne me in the land.
I thought to myself, 'It is well for me that my limbs are
still strong,'
And yet, being old, in my heart I feared I should never
live to return.
The Tartar chieftains shoot so well that the birds are afraid
to fly:
From the risk of their arrows I escaped alive and fled
swiftly home.
Hiding all day and walking all night, I crossed the Great
Desert:[1]
Where clouds are dark and the moon black and the sands
eddy in the wind.
Frightened, I sheltered at the Green Grave,[2] where the
frozen grasses are few:
Stealthily I crossed the Yellow River, at night, on the thin
ice,
Suddenly I heard Han[3] drums and the sound of soldiers
coming:
I went to meet them at the road-side, bowing to them as
they came.
But the moving horsemen did not hear that I spoke the
Han tongue:
Their Captain took me for a Tartar born and had me
bound in chains.

[1] The Gobi Desert.

[2] The grave of Chao-chün, a Chinese girl who in 33 B.C. was "bestowed upon the Khan of the Hsiung-nu as a mark of Imperial regard" (Giles). Hers was the only grave in this desolate district on which grass would grow.

[3] *I.e.*, Chinese.

They are sending me away to the south-east, to a low and swampy land:
No one now will take pity on me: resistance is all in vain.
Thinking of this, my voice chokes and I ask of Heaven above,
Was I spared from death only to spend the rest of my years in sorrow?
My native village of Liang-yüan I shall not see again:
My wife and children in the Tartars' land I have fruitlessly deserted.
When I fell among Tartars and was taken prisoner, I pined for the land of Han:
Now that I am back in the land of Han, they have turned me into a Tartar.
Had I but known what my fate would be, I would not have started home!
For the two lands, so wide apart, are alike in the sorrow they bring.
Tartar prisoners in chains!
Of all the sorrows of all the prisoners mine is the hardest to bear!
Never in the world has so great a wrong befallen the lot of man,—
A Han heart and a Han tongue set in the body of a Turk."

THE CHANCELLOR'S GRAVEL-DRIVE

(A Satire on the Maltreatment of Subordinates)

A Government-bull yoked to a Government-cart!
Moored by the bank of Ch'an River, a barge loaded with gravel.
A single load of gravel,
How many pounds it weighs!
Carrying at dawn, carrying at dusk, what is it all for?
They are carrying it towards the Five Gates,
To the West of the Main Road.
Under the shadow of green laurels they are making a gravel-drive.
For yesterday arrove, newly appointed,
The Assistant Chancellor of the Realm,
And was terribly afraid that the wet and mud
Would dirty his horse's hoofs.
The Chancellor's horse's hoofs
Stepped on the gravel and remained perfectly clean;
But the bull employed in dragging the cart
Was almost sweating blood.
The Assistant Chancellor's business
Is to "save men, govern the country
And harmonize Yin and Yang."
Whether the bull's neck is sore
Need not trouble him at all.

[1] The negative and positive principles in nature.

THE MAN WHO DREAMED OF FAIRIES

This poem is an attack on the Emperor Hsien-tsung, A.D. 806-820, who "was devoted to magic." A Taoist wizard told him that herbs of longevity grew near the city of T'ai-chou. The Emperor at once appointed him prefect of the place, "pour lui permettre d'herboriser plus à son aise" (Wieger, Textes III, 1723). When the censors protested, the Emperor replied: "The ruin of a single district would be a small price to pay, if it could procure longevity for the Lord of Men."

THERE was once a man who dreamt he went to Heaven:
His dream-body soared aloft through space.
He rode on the back of a white-plumed crane,
And was led on his flight by two crimson banners.
Whirring of wings and flapping of coat tails!
Jade bells suddenly all a-tinkle!
Half way to Heaven, he looked down beneath him,
Down on the dark turmoil of the World.
Gradually he lost the place of his native town;
Mountains and water—nothing else distinct.
The Eastern Ocean—a single strip of white:
The Hills of China,—five specks of green.
Gliding past him a host of fairies swept
In long procession to the Palace of the Jade City.
How should he guess that the children of Tzŭ-mēn[1]
Bow to the throne like courtiers of earthly kings?
They take him to the presence of the Mighty Jade Emperor:
He bows his head and proffers loyal homage.
The Emperor says: "We see you have fairy talents:

[1] *I.e.*, the Immortals.

Be of good heart and do not slight yourself.
We shall send to fetch you in fifteen years
And give you a place in the Courtyard of Immortality."
Twice bowing, he acknowledged the gracious words:
Then woke from sleep, full of wonder and joy.
He hid his secret and dared not tell it abroad:
But vowed a vow he would live in a cave of rock.
From love and affection he severed kith and kin:
From his eating and drinking he omitted savoury and spice.
His morning meal was a dish of coral-dust:
At night he sipped an essence of dewy mists.
In the empty mountains he lived for thirty years
Daily watching for the Heavenly Coach to come.
The time of appointment was already long past,
But of wings and coach-bells—still no sound.
His teeth and hair daily withered and decayed:
His ears and eyes gradually lost their keenness.
One morning he suffered the Common Change
And his body was one with the dust and dirt of the hill.
Gods and fairies! If indeed such things there be,
Their ways are beyond the striving of mortal men.
If you have not on your skull the Golden Bump's protrusion,
If your name is absent from the rolls of the Red Terrace,
In vain you learn the "Method of Avoiding Food":
For naught you study the "Book of Alchemic Lore."
Though you sweat and toil, what shall your trouble bring?
You will only shorten the five-score years of your span.
Sad, alas, the man who dreamt of Fairies!
For a single dream spoiled his whole life.

MAGIC

BOUNDLESS, the great sea.
Straight down,—no bottom: sideways,—no border.
Of cloudy waves and misty billows down in the uttermost depths
Men have fabled, in the midst there stand three sacred hills.
On the hills, thick growing,—herbs that banish Death.
Wings grow on those who eat them and they turn into heavenly "hsien."
The Lord of Ch'in[1] and of Wu of Han[2] believed in these stories:
And magic-workers year by year were sent to gather the herbs.
The Blessed Islands, now and of old, what but an empty tale?
The misty waters spread before them and they knew not where to seek.
Boundless, the great sea.
Dauntless, the mighty wind.
Their eyes search but cannot see the shores of the Blessed Islands.
They cannot find the Blessed Isles and yet they dare not return:
Youths and maidens that began the quest grew grey on board the boat.

[1] The "First Emperor," 259-210 B.C.
[2] Wu Ti, 156-87 B.C.

They found that the writings of Hsü Fu[1] were all boasts and lies:
To the Lofty Principle and Great Unity in vain they raised their prayers.
Do you not see
The graves on the top of Black Horse Hill[2] and the tombs at Mo-ling?[3]
What is left but the sighing wind blowing in the tangled grasses?
Yes, and what is more,
The Dark and Primal Master of Sages in his five thousand words[4]
Never spoke of herbs,
Never spoke of "hsien,"
Nor spoke of soaring in broad daylight up to the blue heaven.

THE TWO RED TOWERS

(A SATIRE AGAINST CLERICALISM)

THE Two Red Towers
North and south rise facing each other.
I beg to ask, to whom do they belong?
To the two Princes of the period Chēng Yüan.[5]
The two Princes blew on their flutes and drew down fairies from the sky,

[1] = Hsü Shih. Giles, 1276.
[2] The burial-places of these two Emperors.
[3] *Ibid.*
[4] Lao-tzŭ, in the Tao Tē Ching.
[5] 785-805.

Who carried them off through the Five Clouds, soaring away to Heaven.
Their halls and houses, that they could not take with them,
Were turned into Temples planted in the Dust of the World.
In the tiring-rooms and dancers' towers all is silent and still;
Only the willows like dancers' arms, and the pond like a mirror.
When the flowers are falling at yellow twilight, when things are sad and hushed,
One does not hear songs and flutes, but only chimes and bells.
The Imperial Patent on the Temple doors is written in letters of gold;
For nuns' quarters and monks' cells ample space is allowed.
For green moss and bright moonlight—plenty of room provided;
In a hovel opposite is a sick man who has hardly room to lie down.
I remember once when at P'ing-yang they were building a great man's house
How it swallowed up the housing space of thousands of ordinary men.
The Immortals[1] are leaving us, two by two, and their houses are turned into Temples;
I begin to fear that the whole world will become a vast convent.

[1] Hsien Tsung's brothers?

THE CHARCOAL-SELLER

(A SATIRE AGAINST "KOMMANDATUR")

AN old charcoal-seller
Cutting wood and burning charcoal in the forests of the Southern Mountain.
His face, stained with dust and ashes, has turned to the colour of smoke.
The hair on his temples is streaked with gray: his ten fingers are black.
The money he gets by selling charcoal, how far does it go?
It is just enough to clothe his limbs and put food in his mouth.
Although, alas, the coat on his back is a coat without lining,
He hopes for the coming of cold weather, to send up the price of coal!
Last night, outside the city,—a whole foot of snow;
At dawn he drives the charcoal wagon along the frozen ruts.
Oxen,—weary; man,—hungry: the sun, already high;
Outside the Gate, to the south of the Market, at last they stop in the mud.
Suddenly, a pair of prancing horsemen. Who can it be coming?
A public official in a yellow coat and a boy in a white shirt.
In their hands they hold a written warrant: on their tongues —the words of an order;

They turn back the wagon and curse the oxen, leading them off to the north.
A whole wagon of charcoal,
More than a thousand pieces!
If officials choose to take it away, the woodman may not complain.
Half a piece of red silk and a single yard of damask,
The Courtiers have tied to the oxen's collar, as the price of a wagon of coal!

THE POLITICIAN

I WAS going to the City to sell the herbs I had plucked;
On the way I rested by some trees at the Blue Gate.
Along the road there came a horseman riding;
Whose face was pale with a strange look of dread.
Friends and relations, waiting to say good-bye,
Pressed at his side, but he did not dare to pause.
I, in wonder, asked the people about me
Who he was and what had happened to him.
They told me this was a Privy Councillor
Whose grave duties were like the pivot of State.
His food allowance was ten thousand cash;
Three times a day the Emperor came to his house.
Yesterday he was called to a meeting of Heroes:
To-day he is banished to the country of Yai-chou.
So always, the Counsellors of Kings;
Favour and ruin changed between dawn and dusk!

Green, green,—the grass of the Eastern Suburb;
And amid the grass, a road that leads to the hills.
Resting in peace among the white clouds,
At last he has made a "coup" that cannot fail!

THE OLD MAN WITH THE BROKEN ARM

(A SATIRE ON MILITARISM)

AT Hsin-fēng an old man—four-score and eight;
The hair on his head and the hair of his eyebrows—white as the new snow.
Leaning on the shoulders of his great-grandchildren, he walks in front of the Inn;
With his left arm he leans on their shoulders; his right arm is broken.
I asked the old man how many years had passed since he broke his arm;
I also asked the cause of the injury, how and why it happened?
The old man said he was born and reared in the District of Hsin-fēng;
At the time of his birth—a wise reign; no wars or discords.
"Often I listened in the Pear-Tree Garden to the sound of flute and song;
Naught I knew of banner and lance; nothing of arrow or bow.

Then came the wars of T'ien-pao[1] and the great levy of men;
Of three men in each house,—one man was taken.
And those to whom the lot fell, where were they taken to?
Five months' journey, a thousand miles—away to Yün-nan.
We heard it said that in Yün-nan there flows the Lu River;
As the flowers fall from the pepper-trees, poisonous vapours rise.
When the great army waded across, the water seethed like a cauldron;
When barely ten had entered the water, two or three were dead.
To the north of my village, to the south of my village the sound of weeping and wailing,
Children parting from fathers and mothers; husbands parting from wives.
Everyone says that in expeditions against the Min tribes
Of a million men who are sent out, not one returns.
I, that am old, was then twenty-four;
My name and fore-name were written down in the rolls of the Board of War.
In the depth of the night not daring to let any one know
I secretly took a huge stone and dashed it against my arm.
For drawing the bow and waving the banner now wholly unfit;
I knew henceforward I should not be sent to fight in Yün-nan.
Bones broken and sinews wounded could not fail to hurt;
I was ready enough to bear pain, if only I got back home.
My arm—broken ever since; it was sixty years ago.

[1] A.D. 742-755.

One limb, although destroyed,—whole body safe!
But even now on winter nights when the wind and rain blow
From evening on till day's dawn I cannot sleep for pain.
Not sleeping for pain
Is a small thing to bear,
Compared with the joy of being alive when all the rest are dead.
For otherwise, years ago, at the ford of Lu River
My body would have died and my soul hovered by the bones that no one gathered.
A ghost, I'd have wandered in Yün-nan, always looking for home.
Over the graves of ten thousand soldiers, mournfully hovering."
So the old man spoke,
And I bid you listen to his words
Have you not heard
That the Prime Minister of K'ai-yüan,[1] Sung K'ai-fu,
Did not reward frontier exploits, lest a spirit of aggression should prevail?
And have you not heard
That the Prime Minister of T'ien-Pao, Yang Kuo-chung[2]
Desiring to win imperial favour, started a frontier war?
But long before he could win the war, people had lost their temper;
Ask the man with the broken arm in the village of Hsin-fēng?

[1] 713-742.
[2] Cousin of the notorious mistress of Ming-huang, Yang Kuei-fei.

KEPT WAITING IN THE BOAT AT CHIU-K'OU TEN DAYS BY AN ADVERSE WIND

White billows and huge waves block the river crossing;
Wherever I go, danger and difficulty; whatever I do, failure.
Just as in my worldly career I wander and lose the road,
So when I come to the river crossing, I am stopped by contrary winds.
Of fishes and prawns sodden in the rain the smell fills my nostrils;
With the stings of insects that come with the fog, my whole body is sore.
I am growing old, time flies, and my short span runs out,
While I sit in a boat at Chiu-k'ou, wasting ten days!

ON BOARD SHIP: READING YÜAN CHĒN'S POEMS

I take your poems in my hand and read them beside the candle;
The poems are finished: the candle is low: dawn not yet come.
With sore eyes by the guttering candle still I sit in the dark,
Listening to waves that, driven by the wind, strike the prow of the ship.

ARRIVING AT HSÜN-YANG

(Two Poems)

(1)

A bend of the river brings into view two triumphal arches;
That is the gate in the western wall of the suburbs of Hsün-yang.
I have still to travel in my solitary boat three or four leagues—
By misty waters and rainy sands, while the yellow dusk thickens.

(2)

We are almost come to Hsün-yang: how my thoughts are stirred
As we pass to the south of Yü Liang's[1] tower and the east of P'ēn Port.
The forest trees are leafless and withered,—after the mountain rain;
The roofs of the houses are hidden low among the river mists.
The horses, fed on water grass, are too weak to carry their load;
The cottage walls of wattle and thatch let the wind blow on one's bed.
In the distance I see red-wheeled coaches driving from the town-gate;
They have taken the trouble, these civil people, to meet their new Prefect!

[1] Died A.D. 340. Giles, 2526.

MADLY SINGING IN THE MOUNTAINS

THERE is no one among men that has not a special failing:
And my failing consists in writing verses.
I have broken away from the thousand ties of life:
But this infirmity still remains behind.
Each time that I look at a fine landscape:
Each time that I meet a loved friend,
I raise my voice and recite a stanza of poetry
And am glad as though a God had crossed my path.
Ever since the day I was banished to Hsün-yang
Half my time I have lived among the hills.
And often, when I have finished a new poem,
Alone I climb the road to the Eastern Rock.
I lean my body on the banks of white stone:
I pull down with my hands a green cassia branch.
My mad singing startles the valleys and hills:
The apes and birds all come to peep.
Fearing to become a laughing-stock to the world,
I choose a place that is unfrequented by men.

RELEASING A MIGRANT "YEN" (WILD GOOSE)

At Nine Rivers,[1] in the tenth year,[2] in winter,—heavy snow;
The river-water covered with ice and the forests broken with their load.[3]
The birds of the air, hungry and cold, went flying east and west;
And with them flew a migrant "yen," loudly clamouring for food.
Among the snow it pecked for grass; and rested on the surface of the ice:
It tried with its wings to scale the sky; but its tired flight was slow.
The boys of the river spread a net and caught the bird as it flew;
They took it in their hands to the city-market and sold it there alive.
I that was once a man of the North am now an exile here:
Bird and man, in their different kind, are each strangers in the south.
And because the sight of an exiled bird wounded an exile's heart,
I paid your ransom and set you free, and you flew away to the clouds.

[1] Kiukiang, the poet's place of exile.
[2] A.D. 815. His first winter at Kiukiang.
[3] By the weight of snow.

Yen, Yen, flying to the clouds, tell me, whither shall you go?
Of all things I bid you, do not fly to the land of the north-west
In Huai-hsi there are rebel bands[1] that have not been subdued;
And a thousand thousand armoured men have long been camped in war.
The official army and the rebel army have grown old in their opposite trenches;
The soldier's rations have grown so small, they'll be glad of even you.
The brave boys, in their hungry plight, will shoot you and eat your flesh;
They will pluck from your body those long feathers and make them into arrow-wings!

TO A PORTRAIT PAINTER WHO DESIRED HIM TO SIT

You, so bravely splashing reds and blues!
Just when *I* am getting wrinkled and old.
Why should you waste the moments of inspiration
Tracing the withered limbs of a sick man?
Tall, tall is the Palace of Ch'i-lin;[2]
But my deeds have not been frescoed on its walls.
Minutely limned on a foot of painting silk—
What can I do with a portrait such as *that*?

[1] The revolt of Wu Yüan-chi.

[2] One of the "Record Offices" of the T'ang dynasty, where meritorious deeds were illustrated on the walls.

SEPARATION

YESTERDAY I heard that such-a-one was gone;
This morning they tell me that so-and-so is dead.
Of friends and acquaintances more than two-thirds
Have suffered change and passed to the Land of Ghosts.
Those that are gone I shall not see again;
They, alas, are for ever finished and done.
Those that are left,—where are they now?
They are all scattered,—a thousand miles away.
Those I have known and loved through all my life,
On the fingers of my hand—how many do I count?
Only the prefects of T'ung, Kuo and Li
And Fēng Province—just those four.[1]
Longing for each other we are all grown gray;
Through the Fleeting World rolled like a wave in the stream.
Alas that the feasts and frolics of old days
Have withered and vanished, bringing us to this!
When shall we meet and drink a cup of wine
And laughing gaze into each other's eyes?

[1] Yüan Chēn (d. 831), Ts'ui Hsüan-liang (d. 833), Liu Yü-hsi (d. 842), and Li Chien (d. 821).

HAVING CLIMBED TO THE TOPMOST PEAK OF THE INCENSE-BURNER MOUNTAIN

Up and up, the Incense-burner Peak!
In my heart is stored what my eyes and ears perceived.
All the year—detained by official business;
To-day at last I got a chance to go.
Grasping the creepers, I clung to dangerous rocks;
My hands and feet—weary with groping for hold.
There came with me three or four friends,
But two friends dared not go further.
At last we reached the topmost crest of the Peak;
My eyes were blinded, my soul rocked and reeled.
The chasm beneath me—ten thousand feet;
The ground I stood on, only a foot wide.
If you have not exhausted the scope of seeing and hearing,
How can you realize the wideness of the world?
The waters of the River looked narrow as a ribbon,
P'ēn Castle smaller than a man's fist.
How it clings, the dust of the world's halter!
It chokes my limbs: I cannot shake it away.
Thinking of retirement,[1] I heaved an envious sigh,
Then, with lowered head, came back to the Ants' Nest.

[1] *I.e.*, retirement from office.

EATING BAMBOO-SHOOTS

My new Province is a land of bamboo-groves:
Their shoots in spring fill the valleys and hills.
The mountain woodman cuts an armful of them
And brings them down to sell at the early market.
Things are cheap in proportion as they are common;
For two farthings, I buy a whole bundle.
I put the shoots in a great earthen pot
And heat them up along with boiling rice.
The purple nodules broken,—like an old brocade;
The white skin opened,—like new pearls.
Now every day I eat them recklessly;
For a long time I have not touched meat.
All the time I was living at Lo-yang
They could not give me enough to suit my taste,
Now I can have as many shoots as I please;
For each breath of the south-wind makes a new bamboo!

THE RED COCKATOO

Sent as a present from Annam—
A red cockatoo.
Coloured like the peach-tree blossom,
Speaking with the speech of men.

And they did to it what is always done
To the learned and eloquent.
They took a cage with stout bars
And shut it up inside.

AFTER LUNCH

After lunch—one short nap:
On waking up—two cups of tea.
Raising my head, I see the sun's light
Once again slanting to the south-west.
Those who are happy regret the shortness of the day;
Those who are sad tire of the year's sloth.
But those whose hearts are devoid of joy or sadness
Just go on living, regardless of "short" or "long."

ALARM AT FIRST ENTERING THE YANG-TZE GORGES

Written in 818, when he was being towed up the rapids to Chung-chou.

Above, a mountain ten thousand feet high:
Below, a river a thousand fathoms deep.
A strip of green, walled by cliffs of stone:
Wide enough for the passage of a single reed.[1]
At Chü-t'ang a straight cleft yawns:

1 See Odes, v, 7.

At Yen-yü islands block the stream.
Long before night the walls are black with dusk;
Without wind white waves rise.
The big rocks are like a flat sword:
The little rocks resemble ivory tusks.

We are stuck fast and cannot move a step.
How much the less, three hundred miles?[1]
Frail and slender, the twisted-bamboo rope:
Weak, the dangerous hold of the towers' feet.
A single slip—the whole convoy lost:
And *my* life hangs on *this* thread!
I have heard a saying "He that has an upright heart
Shall walk scathless through the lands of Man and Mo."[2]
How can I believe that since the world began
In every shipwreck none have drowned but rogues?
And how can I, born in evil days[3]
And fresh from failure,[4] ask a kindness of Fate?
Often I fear that these un-talented limbs
Will be laid at last in an un-named grave!

ON BEING REMOVED FROM HSÜN-YANG AND SENT TO CHUNG-CHOU

A remote place in the mountains of Pa (Ssech'uan)

Before this, when I was stationed at Hsün-yang,
Already I regretted the fewness of friends and guests.
Suddenly, suddenly,—bearing a stricken heart

1 The distance to Chung-chou. 2 Dangerous savages.
3 Of civil war. 4 Alluding to his renewed banishment.

I left the gates, with nothing to comfort me.
Henceforward,—relegated to deep seclusion
In a bottomless gorge, flanked by precipitous mountains,
Five months on end the passage of boats is stopped
By the piled billows that toss and leap like colts.
The inhabitants of Pa resemble wild apes;
Fierce and lusty, they fill the mountains and prairies.
Among such as these I cannot hope for friends
And am pleased with anyone who is even remotely human!

PLANTING FLOWERS ON THE EASTERN EMBANKMENT

Written when Governor of Chung-Chou

I TOOK money and bought flowering trees
And planted them out on the bank to the east of the Keep.
I simply bought whatever had most blooms,
Not caring whether peach, apricot, or plum.
A hundred fruits, all mixed up together;
A thousand branches, flowering in due rotation.
Each has its season coming early or late;
But to all alike the fertile soil is kind.
The red flowers hang like a heavy mist;
The white flowers gleam like a fall of snow.
The wandering bees cannot bear to leave them;
The sweet birds also come there to roost.
In front there flows an ever-running stream;
Beneath there is built a little flat terrace.
Sometimes I sweep the flagstones of the terrace;

Sometimes, in the wind, I raise my cup and drink.
The flower-branches screen my head from the sun;
The flower-buds fall down into my lap.
Alone drinking, alone singing my songs
I do not notice that the moon is level with the steps.
The people of Pa do not care for flowers;
All the spring no one has come to look.
But their Governor General, alone with his cup of wine
Sits till evening and will not move from the place!

CHILDREN

Written *circa* 820

My niece, who is six years old, is called "Miss Tortoise";
My daughter of three,—little "Summer Dress."
One is beginning to learn to joke and talk;
The other can already recite poems and songs.
At morning they play clinging about my feet;
At night they sleep pillowed against my dress.
Why, children, did you reach the world so late,
Coming to me just when my years are spent?
Young things draw our feelings to them;
Old people easily give their hearts.
The sweetest vintage at last turns sour;
The full moon in the end begins to wane.
And so with men the bonds of love and affection
Soon may change to a load of sorrow and care.
But all the world is bound by love's ties;
Why did I think that I alone should escape?

PRUNING TREES

TREES growing—right in front of my window;
The trees are high and the leaves grow thick.
Sad alas! the distant mountain view
Obscured by this, dimly shows between.
One morning I took knife and axe;
With my own hand I lopped the branches off.
Ten thousand leaves fall about my head;
A thousand hills came before my eyes.
Suddenly, as when clouds or mists break
And straight through, the blue sky appears;
Again, like the face of a friend one has loved
Seen at last after an age of parting.
First there came a gentle wind blowing;
One by one the birds flew back to the tree.
To ease my mind I gazed to the South East;
As my eyes wandered, my thoughts went far away.
Of men there is none that has not some preference;
Of things there is none but mixes good with ill.
It was not that I did not love the tender branches;
But better still,—to see the green hills!

BEING VISITED BY A FRIEND DURING ILLNESS

I HAVE been ill so long that I do not count the days;
At the southern window, evening—and again evening.
Sadly chirping in the grasses under my eaves
The winter sparrows morning and evening sing.
By an effort I rise and lean heavily on my bed;
Tottering I step towards the door of the courtyard.
By chance I meet a friend who is coming to see me;
Just as if I had gone specially to meet him.
They took my couch and placed it in the setting sun;
They spread my rug and I leaned on the balcony-pillar.
Tranquil talk was better than any medicine;
Gradually the feelings came back to my numbed heart.

ON THE WAY TO HANGCHOW: ANCHORED ON THE RIVER AT NIGHT

LITTLE sleeping and much grieving,—the traveller
Rises at midnight and looks back towards home.
The sands are bright with moonlight that joins the shores;
The sail is white with dew that has covered the boat.
Nearing the sea, the river grows broader and broader:
Approaching autumn,—the nights longer and longer.
Thirty times we have slept amid mists and waves,
And still we have not reached Hang-chow!

STOPPING THE NIGHT AT JUNG-YANG

I GREW up at Jung-yang;
I was still young when I left.
On and on,—forty years passed
Till again I stayed for the night at Jung-yang.
When I went away, I was only eleven or twelve;
This year I am turned fifty-six.
Yet thinking back to the times of my childish games,
Whole and undimmed, still they rise before me.
The old houses have all disappeared;
Down in the village none of my people are left.
It is not only that streets and buildings have changed;
But steep is level and level changed to steep!
Alone unchanged, the waters of Ch'iu and Yu
Passionless,—flow in their old course.

THE SILVER SPOON

While on the road to his new province, Hang-chow, in 822, he sends a silver spoon to his niece A-kuei, whom he had been obliged to leave behind with her nurse, old Mrs. Ts'ao.

To distant service my heart is well accustomed;
When I left home, it wasn't *that* which was difficult
But because I had to leave Miss Kuei at home—
For this it was that tears filled my eyes.
Little girls ought to be daintily fed:
Mrs. Ts'ao, please see to this!
That's why I've packed and sent a silver spoon;
You will think of me and eat up your food nicely!

THE HAT GIVEN TO THE POET BY LI CHIEN

LONG ago to a white-haired gentleman
You made the present of a black gauze hat.
The gauze hat still sits on my head;
But you already are gone to the Nether Springs.
The thing is old, but still fit to wear;
The man is gone and will never be seen again.
Out on the hill the moon is shining to-night
And the trees on your tomb are swayed by the autumn wind.

THE BIG RUG

THAT so many of the poor should suffer from cold what can we do to prevent?
To bring warmth to a single body is not much use.
I wish I had a big rug ten thousand feet long,
Which at one time could cover up every inch of the City.

AFTER GETTING DRUNK, BECOMING SOBER IN THE NIGHT

OUR party scattered at yellow dusk and I came home to bed;
I woke at midnight and went for a walk, leaning heavily on a friend.
As I lay on my pillow my vinous complexion, soothed by sleep, grew sober;
In front of the tower the ocean moon, accompanying the tide, had risen.
The swallows, about to return to the beams, went back to roost again;
The candle at my window, just going out, suddenly revived its light.
All the time till dawn came, still my thoughts were muddled;
And in my ears something sounded like the music of flutes and strings.

REALIZING THE FUTILITY OF LIFE

Written on the wall of a priest's cell, *circa* 828

EVER since the time when I was a lusty boy
Down till now when I am ill and old,
The things I have cared for have been different at different times,

But my being *busy*, *that* has never changed.
Then on the shore,—building sand-pagodas;
Now, at Court, covered with tinkling jade.
This and that,—equally childish games,
Things whose substance passes in a moment of time!
While the hands are busy, the heart cannot understand;
When there are no Scriptures, then Doctrine is sound.[1]
Even should one zealously strive to learn the Way,
That very striving will make one's error more.

RISING LATE AND PLAYING WITH A-TS'UI, AGED TWO

Written in 831

ALL the morning I have lain perversely in bed;
Now at dusk I rise with many yawns.
My warm stove is quick to get ablaze;
At the cold mirror I am slow in doing my hair.
With melted snow I boil fragrant tea;
Seasoned with curds I cook a milk-pudding.
At my sloth and greed there is no one but me to laugh;
My cheerful vigour none but myself knows.
The taste of my wine is mild and works no poison;
The notes of my harp are soft and bring no sadness.
To the Three Joys in the book of Mencius[2]
I have added the fourth of playing with my baby-boy.

[1] This is the teaching of the Dhyana Sect.
[2] "Mencius," bk. vii, pt. i, 20.

ON A BOX CONTAINING HIS OWN WORKS

I BREAK up cypress and make a book-box;
The box well-made,—and the cypress-wood tough.
In it shall be kept what author's works?
The inscription says PO LO-T'IEN.
All my life has been spent in writing books,
From when I was young till now that I am old.
First and last,—seventy whole volumes;
Big and little,—three thousand themes.[1]
Well I know in the end they'll be scattered and lost;
But I cannot bear to see them thrown away
With my own hand I open and shut the locks,
And put it carefully in front of the book-curtain.
I am like Tēng Pai-tao;[2]
But to-day there is not any Wang Ts'an.[3]
All I can do is to divide them among my daughters
To be left by them to give to my grandchildren.

[1] *I.e.*, separate poems, essays, etc.
[2] Who was obliged to abandon his only child on the roadside.
[3] Who rescued a foundling.

ON BEING SIXTY

Addressed to Liu Mēng-tē, who had asked for a poem. He was the same age as Po Chü-i.

BETWEEN thirty and forty, one is distracted by the Five Lusts;
Between seventy and eighty, one is a prey to a hundred diseases.
But from fifty to sixty one is free from all ills;
Calm and still—the heart enjoys rest.
I have put behind me Love and Greed; I have done with Profit and Fame;
I am still short of illness and decay and far from decrepit age.
Strength of limb I still possess to seek the rivers and hills;
Still my heart has spirit enough to listen to flutes and strings.
At leisure I open new wine and taste several cups;
Drunken I recall old poems and sing a whole volume.
Mēng-tē has asked for a poem and herewith I exhort him
Not to complain of three-score, "the time of obedient ears."[1]

[1] Confucius said that it was not till *sixty* that "his ears obeyed him." This age was therefore called "the time of obedient ears."

CLIMBING THE TERRACE OF KUAN-YIN AND LOOKING AT THE CITY

HUNDREDS of houses, thousands of houses,—like a chess-board.
The twelve streets like a field planted with rows of cabbage.
In the distance perceptible, dim, dim—the fire of approaching dawn;
And a single row of stars lying to the west of the Five Gates.

CLIMBING THE LING YING TERRACE AND LOOKING NORTH

MOUNTING on high I begin to realize the smallness of Man's Domain;
Gazing into distance I begin to know the vanity of the Carnal World.
I turn my head and hurry home—back to the Court and Market,
A single grain of rice falling—into the Great Barn.

GOING TO THE MOUNTAINS WITH A LITTLE DANCING GIRL, AGED FIFTEEN

Written when the poet was about sixty-five

Two top-knots not yet plaited into one.
Of thirty years—just beyond half.
You who are really a lady of silks and satins
Are now become my hill and stream companion!
At the spring fountains together we splash and play:
On the lovely trees together we climb and sport.
Her cheeks grow rosy, as she quickens her sleeve-dancing:
Her brows grow sad, as she slows her song's tune.
Don't go singing the Song of the Willow Branches,[1]
When there 's no one here with a heart for you to break!

DREAMING OF YÜAN CHĒN

This was written eight years after Yüan Chēn's death, when Po-Chü-i was sixty-eight.

At night you came and took my hand and we wandered
together in my dream;
When I woke in the morning there was no one to stop the
tears that fell on my handkerchief.

[1] A plaintive love-song, to which Po Chü-i had himself written words.

On the banks of the Ch'ang my aged body three times[1] has passed through sickness;
At Hsien-yang[2] to the grasses on your grave eight times has autumn come.
You lie buried beneath the springs and your bones are mingled with the clay.
I—lodging in the world of men; my hair white as snow.
A-wei and Han-lang[3] both followed in their turn;
Among the shadows of the Terrace of Night did you know them or not?

A DREAM OF MOUNTAINEERING

Written when he was over seventy

At night, in my dream, I stoutly climbed a mountain,
Going out alone with my staff of holly-wood.
A thousand crags, a hundred hundred valleys—
In my dream-journey none were unexplored
And all the while my feet never grew tired
And my step was as strong as in my young days.
Can it be that when the mind travels backward
The body also returns to its old state?
And can it be, as between body and soul,
That the body may languish, while the soul is still strong?
Soul and body—both are vanities:

[1] Since you died.
[2] Near Ch'ang-an, modern Si-ngan-fu.
[3] Affectionate names of Li Chien and Ts'ui Hsüan-liang.

Dreaming and waking—both alike unreal.
In the day my feet are palsied and tottering;
In the night my steps go striding over the hills.
As day and night are divided in equal parts—
Between the two, I *get* as much as I *lose*.

EASE

Congratulating himself on the comforts of his life after his retirement from office. Written *circa* 844.

LINED coat, warm cap and easy felt slippers,
In the little tower, at the low window, sitting over the sunken brazier.
Body at rest, heart at peace; no need to rise early.
I wonder if the courtiers at the Western Capital know of the things, or not?

ON HEARING SOMEONE SING A POEM BY YÜAN CHĒN

Written long after Chēn's death

No new poems his brush will trace:
Even his fame is dead.
His old poems are deep in dust
At the bottom of boxes and cupboards.
Once lately, when someone was singing,
Suddenly I heard a verse—
Before I had time to catch the words
A pain had stabbed my heart.

THE PHILOSOPHERS

Lao-tzǔ

"Those who speak know nothing;
Those who know are silent."
These words, as I am told,
Were spoken by Lao-tzǔ.
If we are to believe that Lao-tzǔ
Was himself *one who knew*,
How comes it that he wrote a book
Of five thousand words?

Chuang-tzǔ, the Monist

Chuang-tzǔ levels all things
And reduces them to the same Monad.
But *I* say that even in their sameness
Difference may be found.
Although in following the promptings of their nature
They display the same tendency,
Yet it seems to me that in some ways
A phœnix is superior to a reptile!

TAOISM AND BUDDHISM

Written shortly before his death

A TRAVELLER came from across the seas
Telling of strange sights.
"In a deep fold of the sea-hills
I saw a terrace and tower.
In the midst there stood a Fairy Temple
With one niche empty.
They all told me this was waiting
For Lo-t'ien to come."

Traveller, I have studied the Empty Gate;[1]
I am no disciple of Fairies
The story you have just told
Is nothing but an idle tale.
The hills of ocean shall never be
Lo-t'ien's home.
When I leave the earth it will be to go
To the Heaven of Bliss Fulfilled.[2]

[1] Buddhism. The poem is quite frivolous, as is shown by his claim to Bodhisattva-hood.

[2] The "tushita" Heaven, where Bodhisattvas wait till it is time for them to appear on earth as Buddhas.

LAST POEM

. . . .

THEY have put my bed beside the unpainted screen;
They have shifted my stove in front of the blue curtain.
I listen to my grandchildren, reading me a book;
I watch the servants, heating up my soup.
With rapid pencil I answer the poems of friends;
I feel in my pockets and pull out medicine-money.
When this superintendence of trifling affairs is done,
I lie back on my pillows and sleep with my face to the South.

THE END